GCSE Chemistry The Workbook

This book is for anyone doing **GCSE Chemistry**.

It's full of **tricky questions**... each one designed to make you **sweat** — because that's the only way you'll get any **better**.

There are questions to see **what facts** you know. There are questions to see how well you can **apply those facts**. And there are questions to see what you know about **how science works**.

It's also got some daft bits in to try and make the whole experience at least vaguely entertaining for you.

What CGP is all about

Our sole aim here at CGP is to produce the highest quality books — carefully written, immaculately presented and dangerously close to being funny.

Then we work our socks off to get them out to you — at the cheapest possible prices.

Contents

Published by Coordination Group Publications Ltd.

From original material by Paddy Gannon and Richard Parsons.

Editors:
Amy Boutal, Ellen Bowness, Tom Cain, Katherine Craig, Sarah Hilton, Kate Houghton, Sharon Keeley, Kate Redmond, Ami Snelling.

Contributors:
Michael Aicken, Mike Bossart, Mike Dagless, Jane Davies, Ian H. Davis, Max Fishel, Rebecca Harvey, Mark Pilkington, Andy Rankin, Sidney Stringer Community School, Paul Warren, Chris Workman.

ISBN: 978 1 84146 643 9

With thanks to Julie Wakeling for the proofreading.

Groovy website: www.cgpbooks.co.uk

Printed by Elanders Hindson Ltd, Newcastle upon Tyne.
Jolly bits of clipart from CorelDRAW®

Text, design, layout and original illustrations © Coordination Group Publications Ltd. 2006
All rights reserved.

Atoms

Q1 Draw a diagram of a **helium atom** in the space provided and label each type of **particle** on your diagram.

Helium has 2 of each type of particle.

Q2 Fill in the blanks to complete these sentences.

a) The number of .... in an atom tells you what element it is.

b) Neutral atoms have a .... charge.

c) A neutral atom has the same number of .... and ....

Q3 **Complete** this table.

Particle	Mass	Charge
Proton	1	
	1	0
Electron		–1

Q4 This question is about an atom of **magnesium**.

An atom of magnesium can be represented by the following symbol: 

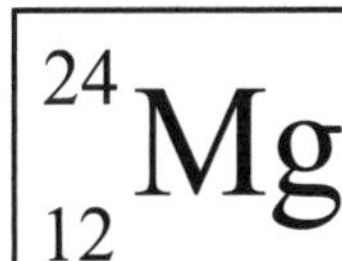

$^{24}_{12}Mg$

a) What is the **mass number** for this atom? What does this number tell you?

....

b) What is the **atomic number** for magnesium? What does this number tell you?

....

c) How many **neutrons** does this atom of magnesium contain? ....

Solids, Liquids and Gases

Q1 For each description below, say whether it refers to the particles of a **solid**, a **liquid** or a **gas**.

a) There are virtually no forces between particles. ....

b) The particles can vibrate but cannot move from place to place. ....

c) The particles can move around but tend to stick together. ....

d) The particles move freely in straight lines. ....

e) There is no fixed volume or shape. ....

f) There is a fixed volume, but no fixed shape. ....

Q2 Choose from the words in the list to fill in the blanks in this paragraph.

"We wanna be free to do what we wanna do."

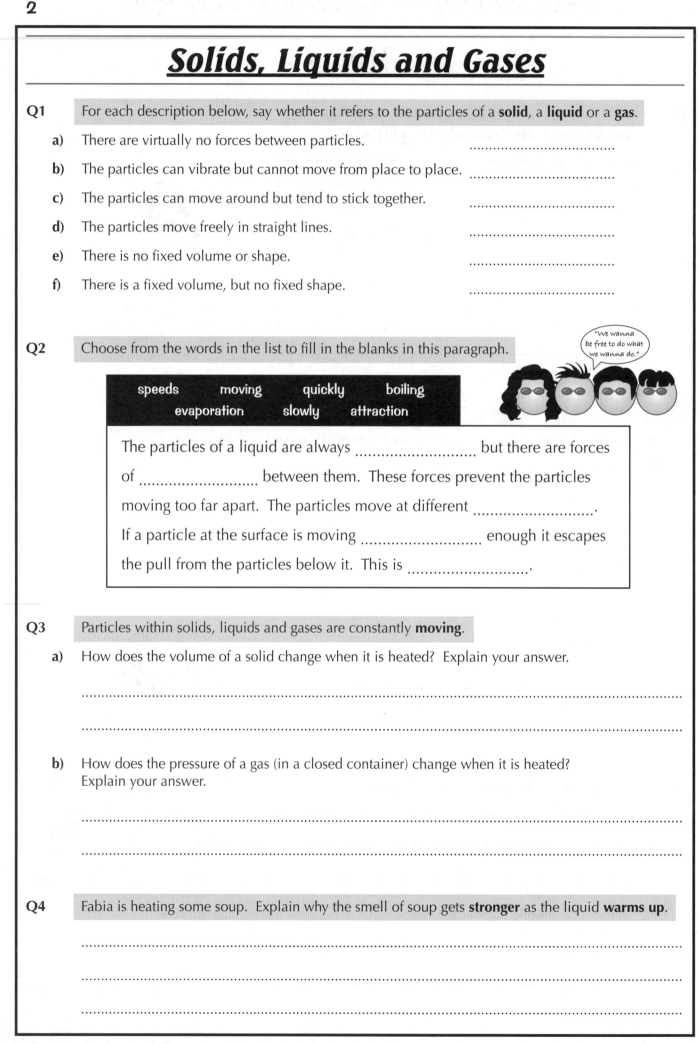

speeds moving quickly boiling evaporation slowly attraction

The particles of a liquid are always .... but there are forces of .... between them. These forces prevent the particles moving too far apart. The particles move at different .... If a particle at the surface is moving .... enough it escapes the pull from the particles below it. This is ....

Q3 Particles within solids, liquids and gases are constantly **moving**.

a) How does the volume of a solid change when it is heated? Explain your answer.

....

....

b) How does the pressure of a gas (in a closed container) change when it is heated? Explain your answer.

....

....

Q4 Fabia is heating some soup. Explain why the smell of soup gets **stronger** as the liquid **warms up**.

....

....

....

Elements, Compounds and Mixtures

Q1 **Sea water** is a **mixture** of water and various dissolved substances, such as sodium chloride (table salt). **Water** is a **compound** of **hydrogen** and **oxygen**.

Tick the correct boxes to show whether these statements are **true** or **false**.

		True	False
a)	The substances in sea water are not chemically bonded to each other.	☐	☐
b)	Water can be separated into hydrogen and oxygen by boiling it.	☐	☐
c)	When sea water is heated until all the water evaporates, the only thing that is left behind is table salt.	☐	☐
d)	The formula for water is H_2O because it contains molecules made up of two hydrogen atoms joined to one oxygen atom.	☐	☐

Q2 Choose from these words to fill in the blanks. Some words may be used more than once.

compounds **taking** **different** **elements** **bonds** **identical** **separate**

Chemicals made up of two or more different elements are called .... They are usually very difficult to .... using physical methods. The properties of compounds are .... to those of the elements used to make them. Mixtures are usually easier to .... because there are no chemical .... between their different parts.

Q3 Look at these diagrams of substances. Circle the ones that contain only **one element**.

copper oxygen water ethane

Q4 Many everyday substances, such as gold and aluminium, are **elements**. Other substances such as water and sugar are not.

Explain what this means in terms of the **atoms** in these substances.

....

....

The Periodic Table

Q1 Choose from these words to fill in the blanks.

left-hand **right-hand** **horizontal** **similar** **different**
vertical **metals** **non-metals** **transition**

a) A group in the periodic table is a .... line of elements.

b) Most of the elements in the periodic table are ....

c) The elements between Group II and Group III are called .... metals.

d) Non-metals are on the .... side of the periodic table.

e) Elements in the same group have .... properties.

Q2 Argon is an extremely **unreactive** gas. Use the periodic table to give the names of two more gases that you would expect to have similar properties to argon.

1. ....

2. ....

Q3 Lithium is **less reactive** than sodium, which is **less reactive** than potassium. Flourine is **more reactive** than chlorine, which is **more reactive** than bromine Use this information to choose the correct words in the sentences below.

a) Reactivity **increases** / **decreases** as you go down Group I.

b) Reactivity **increases** / **decreases** as you go down Group VII.

Have a look at the positions of the elements in the periodic table.

Q4 Tick the correct boxes to show whether the following statements are **true** or **false**.

		True	False
a)	Mendeleev's table of elements had gaps in it.	☐	☐
b)	Mendeleev arranged the elements in order of increasing atomic number.	☐	☐
c)	Undiscovered elements fitted into the gaps in Mendeleev's table.	☐	☐
d)	Elements with similar properties appeared in the same rows.	☐	☐

Balancing Equations

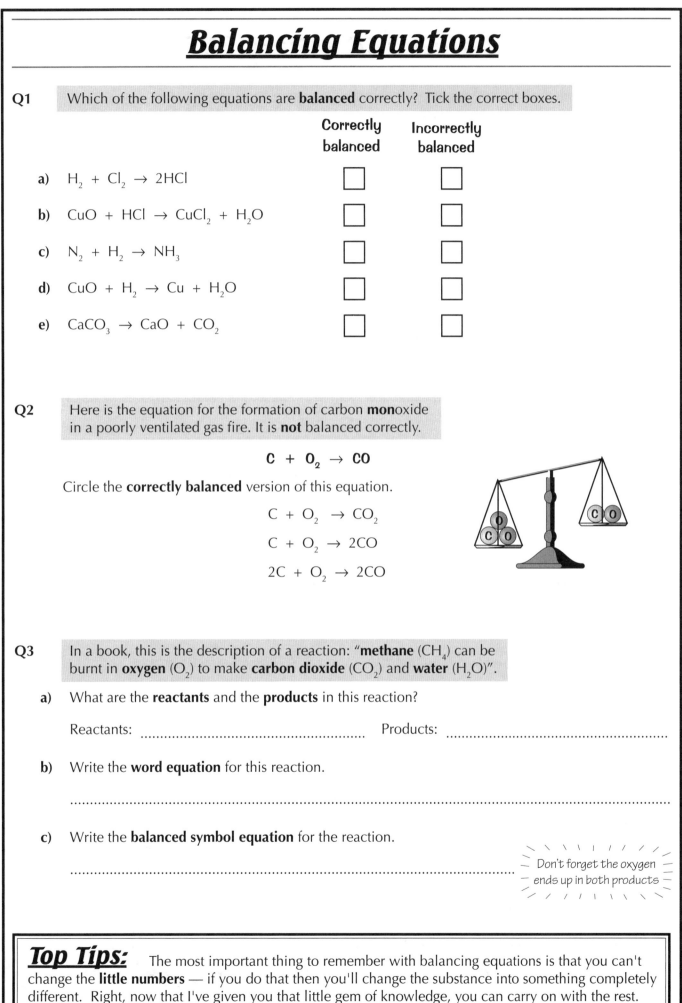

Q1 Which of the following equations are **balanced** correctly? Tick the correct boxes.

		Correctly balanced	Incorrectly balanced
a)	$H_2 + Cl_2 \rightarrow 2HCl$	☐	☐
b)	$CuO + HCl \rightarrow CuCl_2 + H_2O$	☐	☐
c)	$N_2 + H_2 \rightarrow NH_3$	☐	☐
d)	$CuO + H_2 \rightarrow Cu + H_2O$	☐	☐
e)	$CaCO_3 \rightarrow CaO + CO_2$	☐	☐

Q2 Here is the equation for the formation of carbon **mon**oxide in a poorly ventilated gas fire. It is **not** balanced correctly.

$$C + O_2 \rightarrow CO$$

Circle the **correctly balanced** version of this equation.

$C + O_2 \rightarrow CO_2$

$C + O_2 \rightarrow 2CO$

$2C + O_2 \rightarrow 2CO$

Q3 In a book, this is the description of a reaction: "**methane** (CH_4) can be burnt in **oxygen** (O_2) to make **carbon dioxide** (CO_2) and **water** (H_2O)".

a) What are the **reactants** and the **products** in this reaction?

Reactants: .... Products: ....

b) Write the **word equation** for this reaction.

....

c) Write the **balanced symbol equation** for the reaction.

....

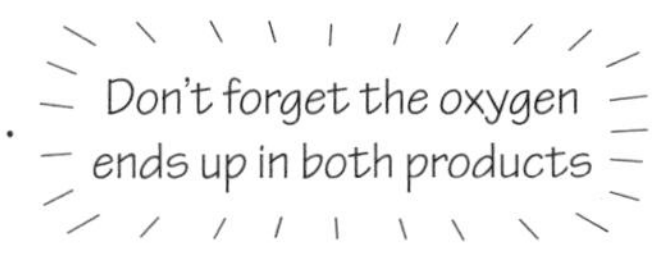

Don't forget the oxygen
ends up in both products

Top Tips: The most important thing to remember with balancing equations is that you can't change the **little numbers** — if you do that then you'll change the substance into something completely different. Right, now that I've given you that little gem of knowledge, you can carry on with the rest.

Balancing Equations

Q4 Write out the balanced **symbol** equations for the picture equations below (some of which are unbalanced).

a)  Na Na + Cl Cl → Na Cl, Na Cl

....

You can draw more pictures to help you balance the unbalanced ones.

b) Li + O O → Li O Li

....

c) Mg O O C O + H Cl → Cl Mg Cl + H O H + O C O

....

d) Li + H O H, H O H → Li O H + H H

....

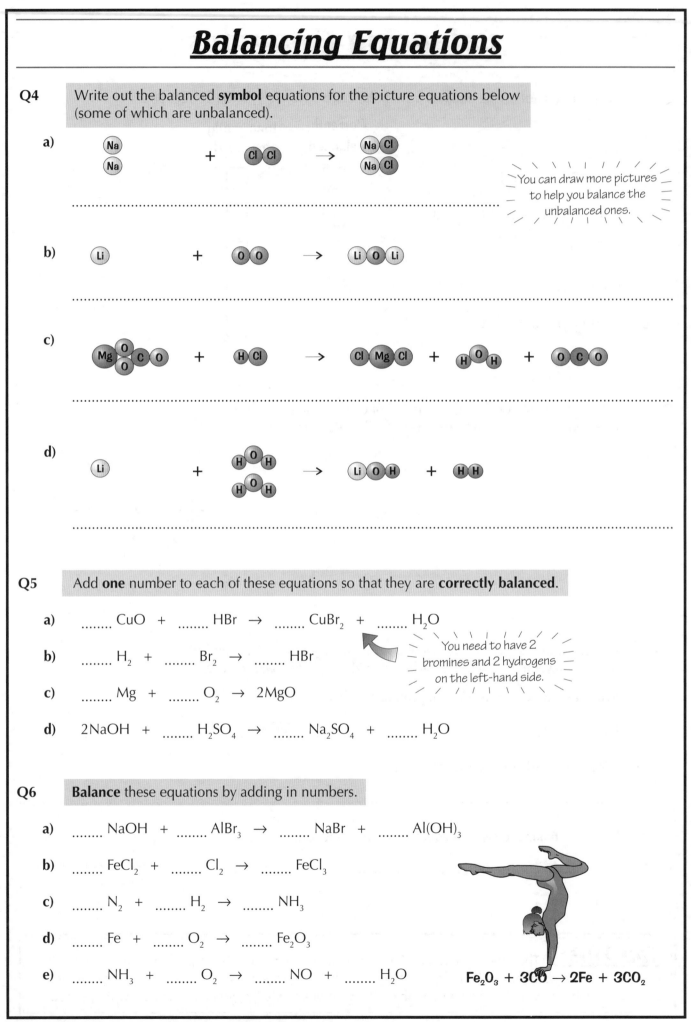

Q5 Add **one** number to each of these equations so that they are **correctly balanced**.

a) .... CuO + ... HBr → .... $CuBr_2$ + ... H_2O

You need to have 2 bromines and 2 hydrogens on the left-hand side.

b) .... H_2 + ... Br_2 → .... HBr

c) .... Mg + ... O_2 → $2MgO$

d) $2NaOH$ + ... H_2SO_4 → .... Na_2SO_4 + ... H_2O

Q6 **Balance** these equations by adding in numbers.

a) .... $NaOH$ + ... $AlBr_3$ → .... $NaBr$ + ... $Al(OH)_3$

b) .... $FeCl_2$ + ... Cl_2 → .... $FeCl_3$

c) .... N_2 + ... H_2 → .... NH_3

d) .... Fe + ... O_2 → .... Fe_2O_3

e) .... NH_3 + ... O_2 → .... NO + ... H_2O

Using Limestone

Q1 Limestone is **mined** in several parts of the UK.

a) What is the chemical name for limestone? ....

b) Limestone can be used to make **quicklime** and **slaked lime**.

i) What is the chemical name for quicklime? ....

ii) Adding water to quicklime makes slaked lime.
Write a balanced symbol equation for this reaction.

....

iii) Describe one use of slaked lime.

....

Q2 Many of the products used to build houses are made with limestone.
Circle the materials that do **not** contain limestone.

cement glass granite paint concrete mortar

Q3 **Carbonates** decompose to form two products.

a) Name the two products formed when limestone is heated and give their formulas.

1. .... 2. ....

b) What **solid** would you expect to be formed when **magnesium carbonate** is heated?

....

c) Write a **symbol equation** for the reaction that occurs when **copper carbonate** ($CuCO_3$) is heated.

....

Q4 Describe three **problems** that **quarrying** limestone can cause.

1. ....

2. ....

3. ....

Metals from Rocks

Q1 This table shows some common **metal ores** and their formulas.

Ore	Formula
Haematite	Fe_2O_3
Magnetite	Fe_3O_4
Pyrites	FeS_2
Galena	PbS
Bauxite	Al_2O_3

a) What is an ore?

....

b) Which two elements are commonly bonded to metals in ores?

....

Q2 Give one **advantage** and one **disadvantage** of mining ores.

....

....

....

Q3 **Copper** is used to make electrical wires.

a) Copper can be extracted from its ore by reduction with carbon.
Why **can't** copper produced in this way be used for electrical wires?

....

b) How is copper that **is** suitable for making electrical wires produced?

....

c) Give another **common use** of copper.

....

Q4 Copper objects can be **recycled**.
Give **two** reasons why it is important to recycle copper.

1. ....

2. ....

Extraction of Metals

Q1 If **zinc** is heated with **copper oxide** this reaction happens:

zinc + copper oxide → copper + zinc oxide

a) Why does this reaction take place? ....

b) i) Would it be possible to produce zinc oxide by reacting zinc with aluminium oxide? ....

ii) Explain your answer. ....

Aluminium is above zinc in the reactivity series.

....

....

Q2 Copper may have been formed when someone accidentally dropped some copper ore into a **wood fire**. When the ashes were cleared away some copper was left.

a) Explain how dropping the ore into the fire led to the extraction of copper.

....

b) Why do you think that copper was one of the first metals to be extracted from its ore?

....

....

Q3 Fill in the blanks in this passage:

> .... can be used to extract metals that are .... it in the reactivity series. Oxygen is removed from the metal oxide in a process called .... Other metals have to be extracted using .... because they are .... reactive.

Q4 Imagine that four new metals, **antium**, **bodium**, **candium** and **dekium** have recently been discovered. Bodium displaces antium but not candium or dekium. Dekium displaces all the others. Put the new metals into their order of reactivity, from the most to the least reactive.

....

Top Tips: Stuff on the reactivity series isn't easy, so don't worry too much if you found these questions difficult. You don't need to learn the reactivity series off by heart, so spend plenty of time making sure that you understand reduction, electrolysis and displacement reactions.

Properties of Metals

Q1 Most **metals** that are used to make everyday objects are found in the **central section** of the periodic table.

a) What name is given to this group of metals?

....

b) What property of a typical metal from this group would make it suitable for electrical wires?

....

Q2 The table below lists some **properties** of two metals, **M** and **N**.

METAL	MELTING POINT (°C)	STRENGTH (MPa)	DENSITY (g/cm^3)	CORROSION RESISTANCE
M	**320**	**100**	**19**	**excellent**
N	**1538**	**350**	**8**	**poor**

a) For each of these applications, say which metal would be most suitable and justify your choice:

i) Girders for making bridges. ....

ii) Protective coatings for other metals. ....

iii) Baking trays for ovens. ....

b) Which property of metal M makes it unsuitable for making lightweight frames for spectacles?

....

Q3 For each of the following **applications** of metals, say which **property** of the metal makes it ideal for the given use. Choose the best answer from the list of typical properties of metals below. You may only use each property **once**.

ductile **malleable** **resists corrosion** **conducts heat**

a) Iron bars are hammered into shape to make horseshoes. ....

b) Copper is used to make the base of saucepans and frying pans. ....

c) Gold is used by dentists to make long-lasting fillings and false teeth. ....

d) Copper is drawn out into thin wires for electrical cables. ....

Q4 What **properties** would you look for if you were asked to choose a **metal** suitable for making knives and forks?

....

....

....

Properties of Metals

Q5 In an experiment some identically sized rods of different materials (A, B, C and D) were **heated** at one end and **temperature sensors** were connected to the other ends. The results of the experiment are shown in the graph.

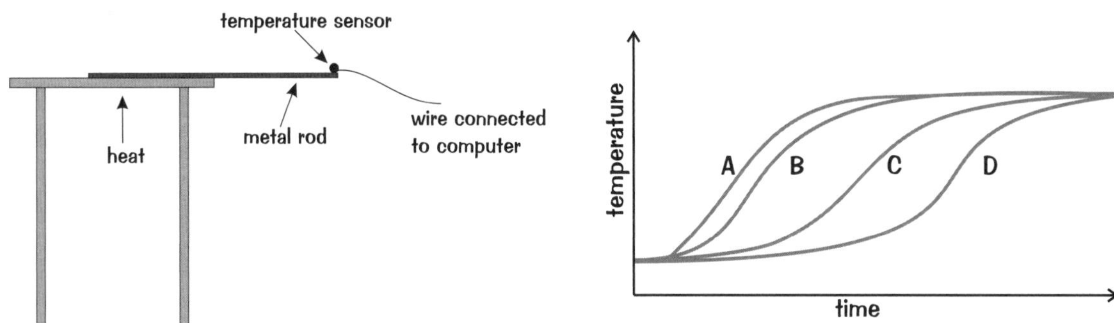

a) Which **two rods** do you think were made from **metals**?

....

b) Which of the metals was the best conductor of heat? How can you tell?

....

Q6 All metals have a similar **structure**. This explains why many of them have similar **properties**.

a) Draw a labelled diagram showing the structure of a typical metal.

b) What is unusual about the electrons in a metal?

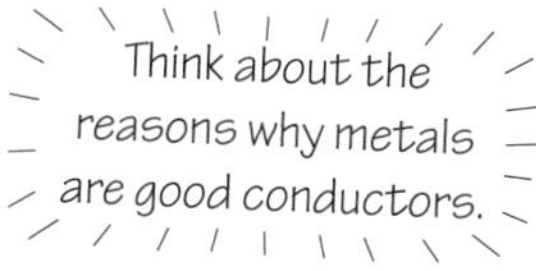

Think about the reasons why metals are good conductors.

....

Q7 Imagine that a space probe has brought a sample of a new element back from Mars. Scientists think that the element is a **metal**, but they aren't certain. Give **three properties** they could look for to provide evidence that the element is a **metal**.

1. ....

2. ....

3. ....

Top Tips: Ever wondered why we don't make bridges out of platinum? Cost is a big factor in the use of metals, so even if a metal is perfect for a job it might not be used because it's too expensive. The cheapest metals are the ones that are both common and easy to extract from their ores.

More Metals

Q1 Complete the following sentences by choosing from the metals below.

gold copper silver nickel titanium

a) Bronze is an alloy that contains ....

b) Cupronickel, which is used in 'silver' coins, contains copper and ....

c) To make gold hard enough for jewellery it is mixed with metals such as ....

Q2 Metals are mixed with other elements to make their **properties** more suitable for their **uses**.

Tick the correct boxes to show whether each statement is **true** or **false**.

		True	False
a)	Aluminium is useful because it does not react with oxygen in the air.	☐	☐
b)	Aluminium is expensive because it has to be extracted using electrolysis.	☐	☐
c)	Impure iron straight from the blast furnace isn't much use as it is too bendy.	☐	☐
d)	Removing all of the impurities produces pure iron which is much harder.	☐	☐
e)	Bronze is an alloy used for making medals and statues.	☐	☐
f)	Pure aluminium isn't suitable for making aircraft as its density is too high.	☐	☐
g)	Cupronickel is a corrosion-resistant alloy used to make coins.	☐	☐

Q3 Most iron is made into the alloy **steel**.

a) What is an **alloy**?

....

....

b) How is **iron** turned into **steel**?

....

....

Tonight Matthew,
I'm going to be...
steel.

More Metals

Q5 Draw a diagram showing the structure of **iron**. Annotate your diagram to explain why iron and other metals can be **bent** and **shaped** without breaking.

Q6 24-carat gold is **pure** gold. 9-carat gold contains **9 parts** gold to 15 parts other metals. 9-carat gold is **harder** and **cheaper** than 24-carat gold.

a) What percentage of 9-carat gold is actually gold?

....

b) Why is 9-carat gold harder than pure gold?

....

....

Q7 Recently, scientists have been developing **smart alloys** with **shape memory** properties.

a) Give an example of a use for shape memory alloys.

....

Smart Alloy of the
Month Award
Presented to: Nitinol
Presented by: CGP

b) What **advantages** do shape memory alloys have over ordinary metals?

....

....

c) Give **two disadvantages** of using smart alloys.

....

....

Top Tips: As you must know by now, metals have lots of pretty useful properties, but they can be made even more useful by being mixed together to make alloys.

Paints and Pigments

Q1 Match each term on the left with the correct meaning on the right.

pigment — holds pigment particles to a surface

colloid — keeps paint runny

solvent — tiny particles dispersed in another material

binding medium — gives paint its colour

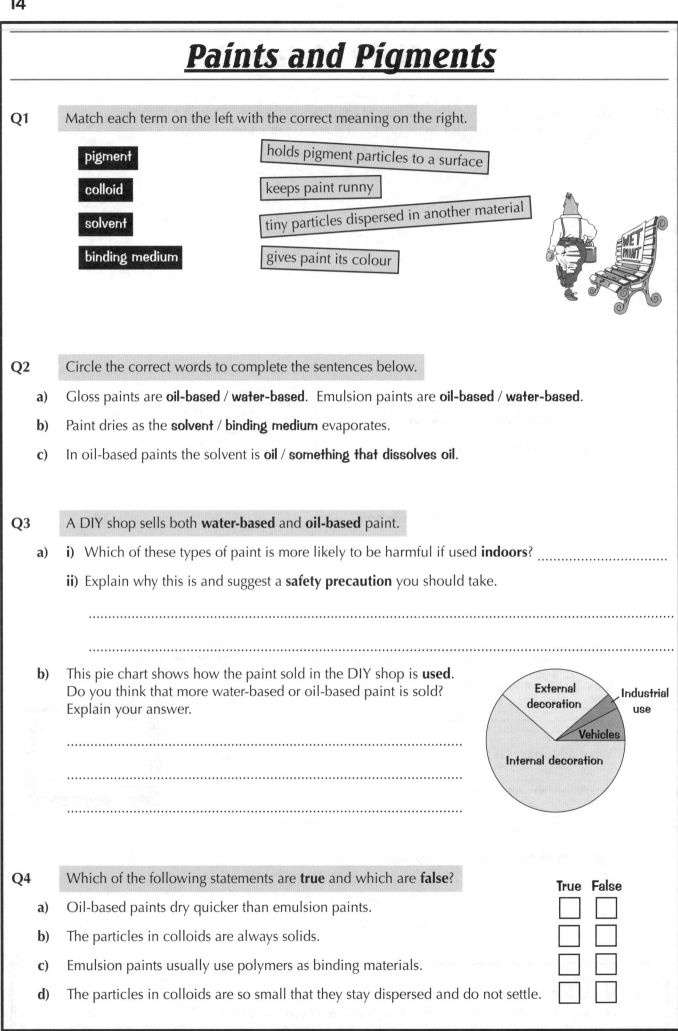

Q2 Circle the correct words to complete the sentences below.

a) Gloss paints are **oil-based** / **water-based**. Emulsion paints are **oil-based** / **water-based**.

b) Paint dries as the **solvent** / **binding medium** evaporates.

c) In oil-based paints the solvent is **oil** / **something that dissolves oil**.

Q3 A DIY shop sells both **water-based** and **oil-based** paint.

a) i) Which of these types of paint is more likely to be harmful if used **indoors**? ....

ii) Explain why this is and suggest a **safety precaution** you should take.

....

....

b) This pie chart shows how the paint sold in the DIY shop is **used**. Do you think that more water-based or oil-based paint is sold? Explain your answer.

....

....

....

External decoration
Industrial use
Vehicles
Internal decoration

Q4 Which of the following statements are **true** and which are **false**?

		True	False
a)	Oil-based paints dry quicker than emulsion paints.	☐	☐
b)	The particles in colloids are always solids.	☐	☐
c)	Emulsion paints usually use polymers as binding materials.	☐	☐
d)	The particles in colloids are so small that they stay dispersed and do not settle.	☐	☐

Fractional Distillation of Crude Oil

Q1 Circle the correct words to complete these sentences.

a) Crude oil is a **mixture** / **compound** of different molecules.

b) The molecules in crude oil **are** / **aren't** chemically bonded to each other.

c) If crude oil were heated, the **first** / **last** thing to boil off would be lubricating oil.

d) Diesel has **larger** / **smaller** molecules than petrol.

Q2 Label this diagram of a **fractionating column** to show where these substances can be collected.

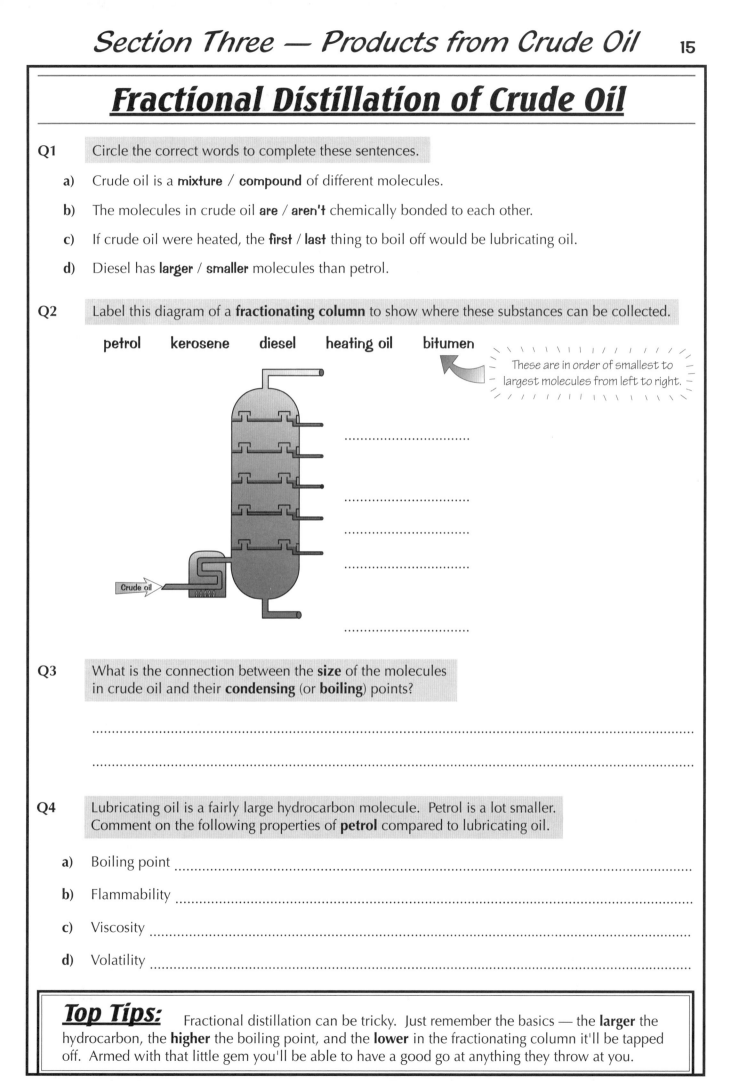

petrol **kerosene** **diesel** **heating oil** **bitumen**

These are in order of smallest to largest molecules from left to right.

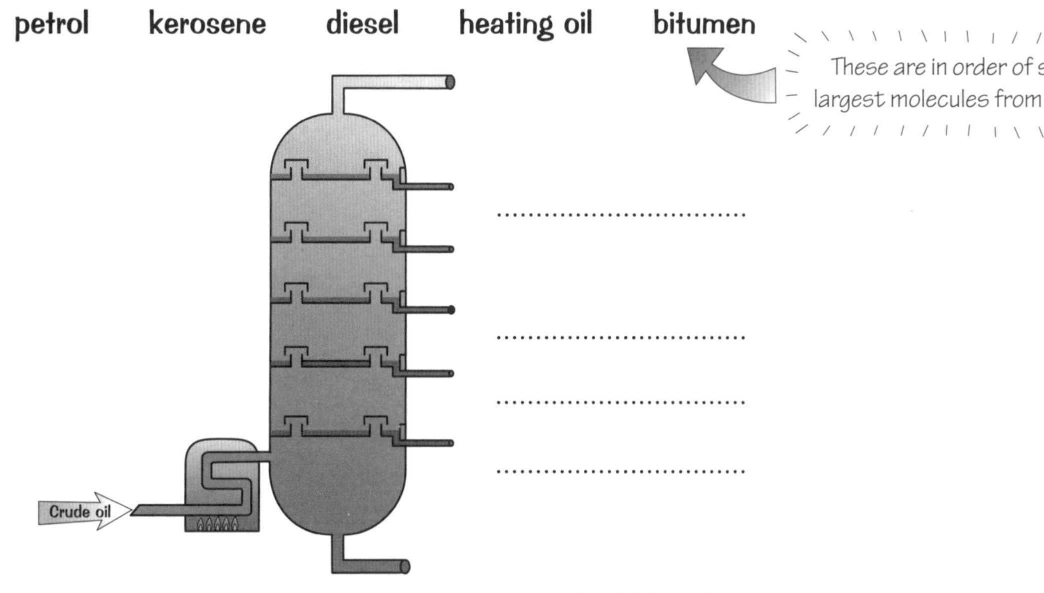

Q3 What is the connection between the **size** of the molecules in crude oil and their **condensing** (or **boiling**) points?

....

....

Q4 Lubricating oil is a fairly large hydrocarbon molecule. Petrol is a lot smaller. Comment on the following properties of **petrol** compared to lubricating oil.

a) Boiling point ....

b) Flammability ....

c) Viscosity ....

d) Volatility ....

Top Tips: Fractional distillation can be tricky. Just remember the basics — the **larger** the hydrocarbon, the **higher** the boiling point, and the **lower** in the fractionating column it'll be tapped off. Armed with that little gem you'll be able to have a good go at anything they throw at you.

Burning Hydrocarbons

Q1 Hydrocarbons make good **fuels**.

a) Which two elements are hydrocarbons made up of? ....

b) Complete the blanks in this definition of a fuel.

Fuels are substances that react with .... to release ....

c) What could cause **incomplete combustion** of a fuel?

....

Today's Lecture:
'My favourite few Ls'

Q2 Burning hydrocarbons in the open air will result in **complete combustion**.

a) What colour **flame** would you see during complete combustion? ....

b) Write a **word equation** for burning a hydrocarbon in the open air.

....

c) Write **balanced symbol equations** for burning these alkanes in open air:

i) Methane: CH_4 + ... → .... + ...

ii) Propane: C_3H_8 + ... → .... + ...

d) Describe how the apparatus on the right could be used to show which substances are produced when hexane (a hydrocarbon) undergoes complete combustion.

....

....

....

....

Q3 **Incomplete combustion** can cause problems.

a) Fill in the blanks to complete the symbol equation for the incomplete combustion of butane.

C_4H_{10} + ... → ...... H_2O + CO_2 + ... + ...

b) Why is incomplete combustion:

i) dangerous? ....

ii) a waste of fuel? ....

iii) messy? ....

c) How might you tell from the **flame** produced that combustion is incomplete?

....

Using Crude Oil as a Fuel

Q1 Crude oil **fractions** are often used as **fuels**.

Remember fuels aren't just used in vehicles.

Give **three** examples of fuels that are made from crude oil.

....

Q2 As crude oil is a **non-renewable** resource, people are keen to find **alternative** energy sources. Suggest a problem with each of these alternative energy sources.

a) **Solar** energy for cars:

....

b) **Wind** energy to power an oven:

....

c) **Nuclear** energy for buses:

....

Q3 Using oil products as fuels causes some **environmental** problems. Outline the environmental problems that are associated with each of the following:

phwoar... nice tank, love

a) **Transporting** crude oil across the sea in tankers.

....

b) **Burning** oil products to release the energy they contain.

....

Q4 Forty years ago some scientists predicted that there would be no oil left by the year 2000, but obviously they were **wrong**. One reason is that modern engines are more **efficient** than ones in the past, so they use less fuel. Give two other reasons why the scientists' prediction was wrong.

....

....

Q5 Write a short paragraph summarising why crude oil is the most **common source** of fuel even though **alternatives** are available.

....

....

....

Alkanes and Alkenes

Q1 Draw the structures of the first four **alkanes** and name each alkane you have drawn.

Q2 a) What is the **general formula** for **alkanes**?

....

If you can't remember it you can work it out by looking at the diagrams you have drawn at the top of the page.

b) **Eicosane** is a hydrocarbon that can be used to make candles. Each molecule of eicosane contains **20 carbon** atoms. What is the **chemical formula** for eicosane?

....

Q3 Complete this table showing the molecular and displayed formulas of some **alkenes**.

Alkene	Formula	Displayed formula
Ethene	a)	b)
c)	C_3H_6	d)

Q4 The general formula for alkenes is $\mathbf{C_nH_{2n}}$. Use it to write down the formulas of these alkenes.

a) pentene (5 carbons) ....

b) hexene (6 carbons) ....

c) octene (8 carbons) ....

d) dodecene (12 carbons) ....

Q5 Tick the boxes to show whether the following are **true** or **false**.

		True	False
a)	Alkenes have double bonds between the hydrogen atoms.	☐	☐
b)	Alkenes are unsaturated.	☐	☐
c)	Alkanes decolourise bromine water.	☐	☐
d)	Alkenes tend to burn cleanly, producing carbon dioxide and water.	☐	☐
e)	Alkenes can form polymers.	☐	☐

Bromine Water

Cracking Crude Oil

Q1 Fill in the gaps with the words below.

high	shorter	long	catalyst	cracking	diesel	molecules	petrol

There is more need for .... chain fractions of crude oil such as .... than for longer chains such as ....

Heating .... hydrocarbon molecules to .... temperatures with a .... breaks them down into smaller .... This is called ....

Q2 Diesel is **cracked** to produce products that are more in demand.

a) Name two of the useful substances that are produced when diesel is cracked.

....

b) Suggest a reason why long hydrocarbons do not make good fuels.

....

c) What type of reaction is cracking? ....

Q3 Put the steps of the cracking process in the **correct order** by writing numbers in the boxes.

☐ The vapour is passed over a catalyst at a high temperature.

☐ The long-chain molecules are heated.

☐ The molecules are cracked on the surface of the catalyst.

☐ They are vaporised (turned into a gas).

Q4 Change this diagram into a **word equation** and a **symbol equation**.

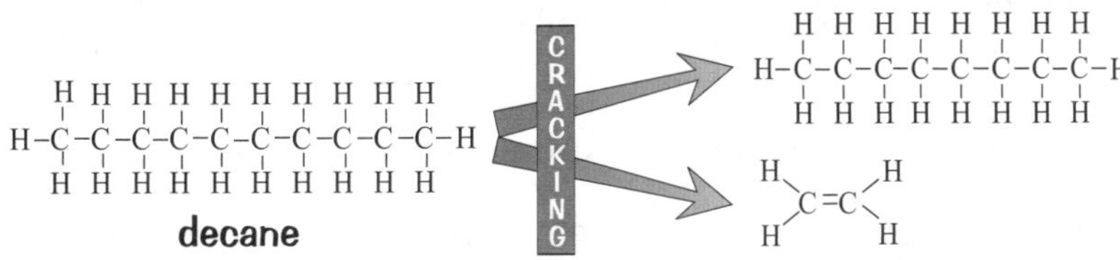

a) Word equation: .... → .... + ...

b) Symbol equation: .... → .... + ...

Making Polymers

Q1 Tick the box next to the **true** statement below.

- ☐ The monomer of polyethene is ethane.
- ☐ The polymer of polyethene is ethane.
- ☐ The monomer of polyethene is ethene.
- ☐ The polymer of polyethane is ethene.

Q2 **Addition polymers** are formed when **unsaturated monomers** link together. Special conditions are needed to make this happen.

a) Explain what feature of the monomer molecules makes them **unsaturated**.

....

b) What **condition** and **substance** are required for the reaction to take place?

....

Q3 The diagram below shows the polymerisation of ethene to form **polyethene**.

$$n\begin{pmatrix} H & & H \\ | & & | \\ C & = & C \\ | & & | \\ H & & H \end{pmatrix} \longrightarrow \left(\begin{matrix} H & & H \\ | & & | \\ -C & - & C- \\ | & & | \\ H & & H \end{matrix}\right)_n$$

many ethene molecules → polyethene

a) Draw a similar diagram in the box below to show the polymerisation of **propene** (C_3H_6).

It's easier if you think of propene as

$$\begin{matrix} H & & & & H \\ & \diagdown & & \diagup & \\ & & C=C & & \\ & \diagup & & \diagdown & \\ H & & & & CH_3 \end{matrix}$$

b) Name the polymer you have drawn ....

Q4 Two rulers, made from **different plastics**, were investigated by bending and heating them. The results are shown in the table.

	RESULT ON BENDING	RESULT ON HEATING
Ruler 1	Ruler bends easily and springs back into shape	Ruler becomes soft and then melts
Ruler 2	Ruler snaps in two	Ruler doesn't soften and eventually turns black

a) Which ruler is made from a polymer that has **strong** forces between its molecules? ....

b) The atoms in both types of plastic are held together with the same strong covalent bonds. Explain why one type of plastic melts and bends more easily than the other.

....

....

Uses of Polymers

Q1 From the list below underline any **properties** you think it is important for a plastic to have if it is to be used to make **Wellington boots**.

low melting point **waterproof** **rigid** **lightweight** **heat resistant**

Q2 This question is about how to dispose of **non-biodegradable** plastics.

a) Lots of plastic is buried in landfill sites. Name one problem with this method.

....

b) Another disposal method is to burn waste plastic. Why isn't this always safe?

....

c) Recycling plastics avoids the problems of disposal. What is the main problem with this solution?

....

Q3 Kate has three black jackets. One is made from **nylon**, another from nylon coated with **polyurethane**, and the third from **a breathable fabric**.

a) Explain why the jacket coated with polyurethane would be better for Kate to wear on a rainy day than the plain nylon jacket.

....

b) Which of the jackets would you advise Kate to take for a week's hiking in Wales? Explain why.

....

....

Q4 Complete the table to show a possible **use** of each polymer given using the options below.

carrier bags **kettles** **window frames** **disposable cups**

POLYMER	PROPERTY	USE
polypropene	heat-resistant	a)
polystyrene foam	thermal insulator	b)
low density polyethene	lightweight	c)
PVC	strong and durable	d)

Top Tips: Polymers have loads of uses, but disposing of them can be a bit of a problem. Many polymers aren't biodegradable — they don't rot so could stay in landfill sites for years.

Chemicals and Food

Q1 When meat or eggs are cooked the **protein** molecules they contain are **denatured**.

a) Explain how cooking causes this chemical change to proteins.

....

....

b) Why does denaturing proteins in meat and eggs make them more appealing to eat?

....

Q2 Fill in the blanks in the passage below choosing from the words in the list to explain why cooking **potatoes** makes them better to eat.

protein   cellulose   carbohydrate   heat   digest   swallow   water

Potatoes are a good source of .... Each potato cell is surrounded by a .... cell wall, which humans can't .... When the potato is cooked, the .... breaks down the cell wall.

Q3 **Lecithin** is added to chocolate drinks to prevent the oils separating out from the water. The diagram shows a molecule of lecithin.

a) Label the **hydrophilic** part and the **hydrophobic** part of the lecithin molecule.

b) Explain how this molecule keeps the oil and water parts of chocolate drinks from separating into two different layers. Include a diagram to help explain your answer.

....

....

....

....

....

....

Top Tips: Hopefully you haven't found these questions too hard. The only really tricky thing in the food topic is emulsifiers — as long as you've got your head around those everything should be OK.

Packaging and Smart Materials

Q1 **Smart materials** can change their properties depending on the external conditions. Give **one possible use** of each of the smart materials described below.

a) A dye that changes from red to green when it is cooled below a certain temperature.

....

b) A material that expands when an electric current is passed through it and produces electricity when squeezed.

....

c) A dye that becomes more transparent as the light intensity decreases.

....

Q2 Food '**goes off**' if it's stored for a long time.

a) How does drying food help to stop this?

....

b) Describe an active packaging method for fresh food that reduces the amount of water present.

....

Q3 Ali buys some freshly cooked chicken with **intelligent packaging**. She records what the dot on the packaging looks like every 12 hours for three days. Her results are shown in the table.

No. hours since the chicken was bought	0	12	24	36	48	60	72
Appearance of dot	◎	◎	◎	◎	◎	◎	◉

a) Describe how this type of intelligent packaging works.

....

....

b) Is this chicken safe to eat after 2 days?

....

Key:

Appearance of dot	Description of food
◎	Very fresh
◎	Still fresh
◎	Still fresh, eat now
◉	Not fresh

c) Ali says "All cooked chicken is safe to eat for up to 60 hours." Give **two things** that are wrong with this statement.

....

....

Food Additives

Q1 Some food additives are used to help prevent foods **'going off'**.

a) Which element in air can cause foods to 'go off'? ....

b) What kind of additives can be used to stop this problem? ....

Q2 Some people avoid certain **additives**.

a) Name an additive that vegetarians would not eat. ....

b) Why are some people unable to eat tartrazine? ....

Q3 Food colourings are usually made up of several different dyes. These can be separated out.

a) What is the name of the **separation** technique that allows us to examine the dyes used in foods?

....

b) Which dye is **more soluble**, A or B? ....

A

B

Q4 Research into the effect of a **food additive X** on the incidence of **migraines** was carried out by monitoring a group of 100 men and 100 women who took a small dose of substance X every day. A control group of 200 people was also monitored.

	% suffering one or more migraines	
	Group taking X	**Control**
Males	11	3
Females	5	6

a) Explain what the conditions would be for the control group and why this group is needed.

....

....

b) What would increase the reliability of the results for this type of research?

....

c) What does the research show for the male and for the female groups?

....

....

d) In such research, **'a correlation or link does not imply a cause'**. Explain what this means.

....

....

Plant Oils in Food

Q1 Oil can be extracted from some **fruits** and **seeds**.

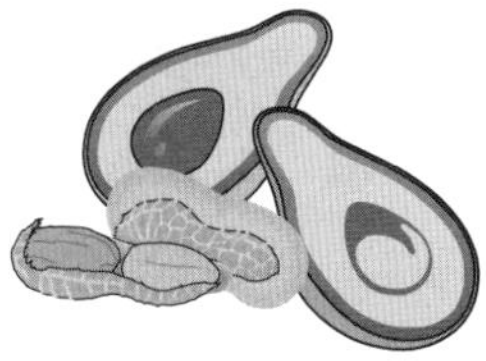

a) Name two fruits and two seeds which are good sources of oil.

Fruits: .... and ....

Seeds: .... and ....

b) Give two uses of plant oils. ....

c) Why is the use of **high pressure** an important part of the oil extraction process?

....

Q2 Each diagram shows part of a fat structure. Draw lines to match each label below to its correct structure.

Saturated animal fat

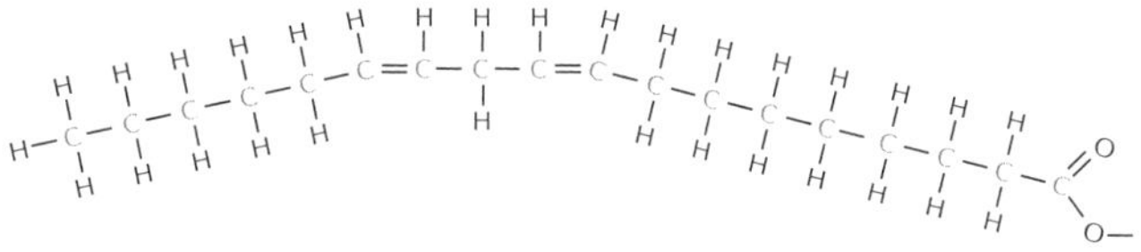

Polyunsaturated grape seed oil

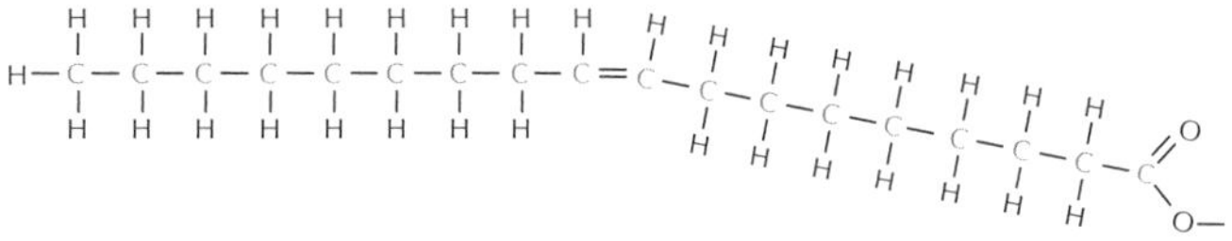

Monounsaturated olive oil

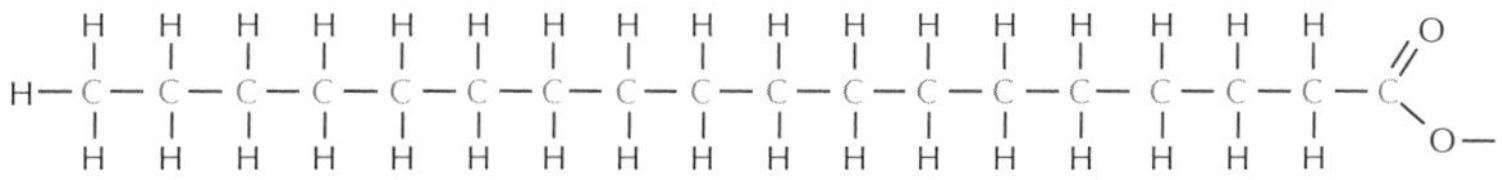

Q3 **Margarine** is usually made from partially **hydrogenated** vegetable oil.

a) Describe the process of hydrogenation.

....

....

b) How does hydrogenation affect the melting points of vegetable oils?

....

Q4 Complete the passage choosing from the words below.

less **decrease** **unsaturated** **cholesterol** **saturated** **more** **increase**

Vegetable oils are usually .... — they contain double bonds in their carbon chains. Animal fats tend to be .... — they contain no double bonds in their carbon chains. In general saturated fats are .... healthy than unsaturated fats as saturated fats .... the amount of .... in the blood.

Plant Oils as Fuel

Q1 **Vegetable oils** can be turned into **fuels**.

a) Name two vegetable oils that can be turned into fuels.

.... and ....

b) What makes vegetable oils suitable for processing into fuels?

....

Q2 **Biodiesel** is a fuel made from vegetable oil. A litre of biodiesel releases **90%** of the energy released by a litre of normal diesel when it is burned.

Normal diesel releases 37 megajoules (37 000 000 J) of energy per litre. How much energy does a litre of biodiesel produce?

....

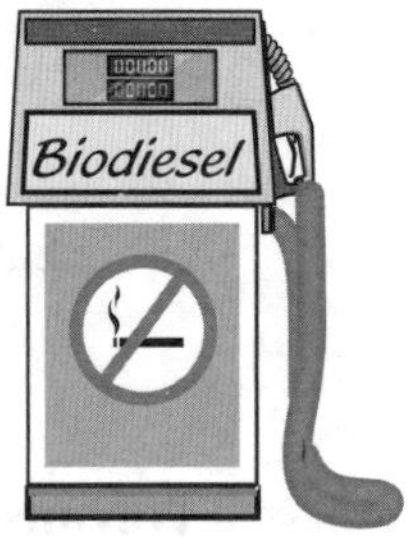

Q3 **Biodiesel** is more **environmentally friendly** than normal diesel or petrol. However, it is unlikely to replace them in the near future.

a) Give three reasons why biodiesel is more environmentally friendly than petrol or normal diesel.

1. ....

2. ....

3. ....

b) Explain why biodiesel is unlikely to replace petrol or normal diesel in the UK in the near future.

....

c) Give two advantages that biodiesel has over other 'green' fuels such as biogas.

1. ....

2. ....

Q4 Biodiesel is said to be **carbon neutral**.

a) Explain why this is.

....

....

b) Why is normal diesel not carbon neutral?

....

....

Plant Oils as Fuel

Q5 Read this passage and answer the questions below.

Biodiesel is a liquid fuel which can be made from vegetable oils. It's renewable, and can be used instead of ordinary diesel in cars. It can also be blended with normal diesel — this is common in some countries, such as France. You don't have to modify your car's engine to use biodiesel.

Biodiesel has several advantages. Producing and using it releases 80% less carbon dioxide overall than producing and using fossil-fuel diesel. So if we want to do something about climate change, using biodiesel would be a good start. Biodiesel is also less harmful if it's accidentally spilled, because it's readily biodegradable.

In the UK, we make most of our biodiesel from recycled cooking oils. But we don't make very much yet — you can only buy it from about 100 filling stations. The Government has been making some effort to encourage us to use more biodiesel. There's one major problem — it's about twice as expensive to make as ordinary diesel.

Most of the price you pay for petrol or diesel is not the cost of the fuel — it's tax, which goes straight to the Government. Over the last decade, the Government has increased fuel taxes, making petrol and diesel more expensive to buy. Part of the reason they've done this is to try to put us off buying them — because burning fossil fuels releases harmful pollutants and contributes to climate change.

So, to make biodiesel cheaper, in 2002, the Government cut the tax rate on it. The tax on biodiesel is now 20p/litre less than it is on normal diesel. This makes biodiesel a similar price to normal diesel. If the Government cuts the tax even further, then more people would be keen to use biodiesel, and more filling stations would start to sell it.

Pay less tax — buy biodiesel

a) In the UK, what do we produce most of our biodiesel from at present?

....

b) What would the environmental impact be if biodiesel were more widely used?

....

....

c) What has the Government done to encourage people to switch from normal diesel to biodiesel?

....

d) If lots more people start buying biodiesel instead of normal diesel, what problem is this likely to cause for the Government?

....

....

e) "I don't want to change to biodiesel. I don't want all the hassle of getting my car modified, and biodiesel costs more. It's just another way for the Government to get money off the taxpayers."

Write a response to this using information from the passage above.

....

....

Ethanol

Q1 What conditions are needed to produce ethanol by reacting **steam** with **ethene**?

....

Q2 Use the words in the box to fill in the gaps in the passage below about **fermentation**.

Each word may be used once, more than once, or not at all.

oxygen	high	low	sugars	enzymes	ethanoic acid	
concentration	10–20%	70 °C	30 °C	50 °C	30–40%	temperature

Fermentation is used to turn .... into ethanol. The reaction happens due to .... found in yeast. The .... needs to be carefully controlled during the reaction — if it is too .... the reaction is very slow, and if it is too .... the .... are destroyed. The optimum is about .... It is also important to stop .... getting into the reaction mixture.

The reaction stops when the .... of the ethanol reaches about ...., because the yeast are killed by the alcohol.

Q3 Choose option **A**, **B** or **C** in each case to answer the questions below about **fermentation**.

a) Which of the following is the correct equation for the production of ethanol?

A $C_6H_{12}O_6 \rightarrow 2C_2H_5OH + 2CO_2$

B $C_2H_{12}O_6 \rightarrow 2C_2H_5OH + 2O_2$

C $C_4H_{12}O_6 \rightarrow 2C_2H_5OH + H_2O$

Answer: ....

b) What is the name of the enzyme in yeast?

A ethanoate **B** fermentase **C** zymase **Answer:** ....

Q4 Ethanol can be mixed with petrol to give a **fuel** called **gasohol**.

a) Name a **country** where gasohol is frequently used and explain why the country is **well suited** to producing gasohol.

Country .... because ....

b) Why is using gasohol more environmentally friendly than using ordinary petrol?

....

....

Perfumes

Q1 New perfumes are sometimes tested on **animals**.

a) Give one reason for testing cosmetic products on animals.

....

b) Give one argument against testing cosmetic products on animals.

....

Q2 A chemist was asked by an **aftershave company** to make some new scents. Her new compounds were then **tested** to see if they were suitable for use in aftershaves. The results of the tests are summarised in the table.

Liquid	Boiling point (°C)	Does it dissolve in water?	Does it react with water?
A	85	no	yes
B	75	yes	no
C	80	no	no
D	150	no	no

a) Which one of these would you use as the scent in an aftershave and why?

....

b) Give two further tests that should be carried out before the chemical can be used in the aftershave.

....

Q3 Below are a set of instructions for reacting **ethanoic acid** with **ethanol** to form the ester **ethyl ethanoate**.

a) Write numbers in the boxes to put the instructions into the right order.

☐ Warm the flask gently on an electric heating plate for 10 minutes.

☐ Put 15 cm^3 of ethanoic acid into a 100 cm^3 conical flask.

☐ When the flask is cool enough to handle, pour its contents into a 250 cm^3 beaker containing 100 cm^3 of sodium carbonate solution.

☐ Add 15 cm^3 of ethanol and a few drops of concentrated sulfuric acid.

☐ Turn off the heat.

b) What is the purpose of the sodium carbonate solution?

....

c) What is the purpose of the sulfuric acid?

....

The Earth's Structure

Q1 Draw a simple diagram of the **Earth's structure** in the space below.
Label the **crust**, **mantle** and **core** and write a brief description of each.

Q2 The map below on the left shows where most of the world's **earthquakes** take place.

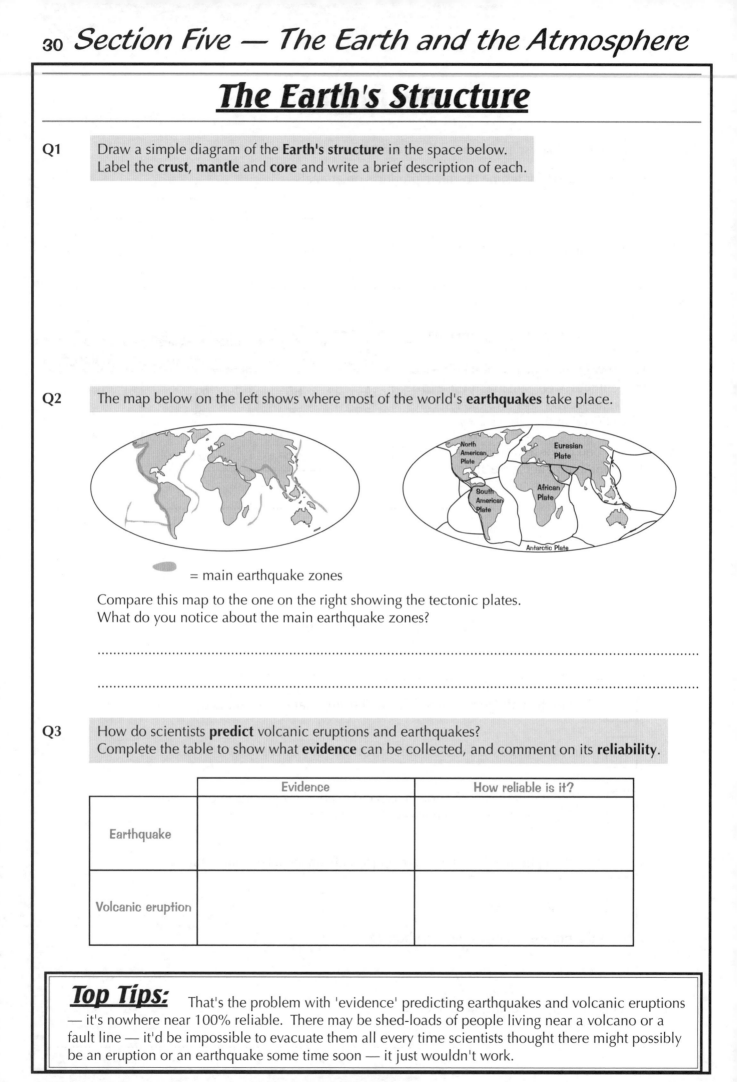

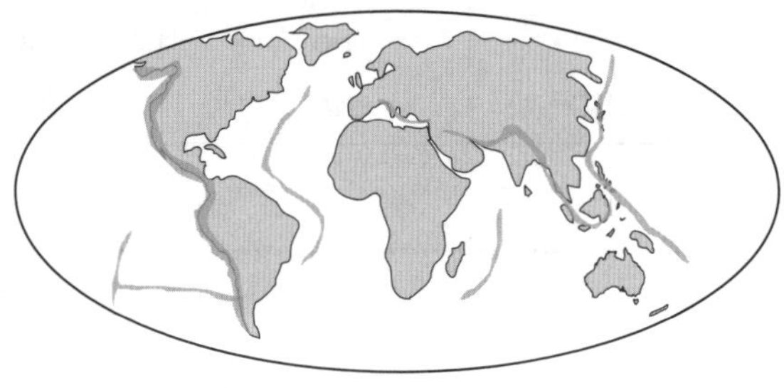

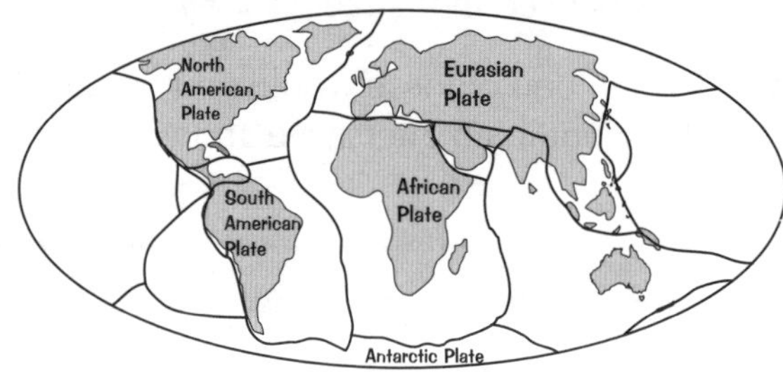

= main earthquake zones

Compare this map to the one on the right showing the tectonic plates.
What do you notice about the main earthquake zones?

....

....

Q3 How do scientists **predict** volcanic eruptions and earthquakes?
Complete the table to show what **evidence** can be collected, and comment on its **reliability**.

	Evidence	How reliable is it?
Earthquake		
Volcanic eruption		

Top Tips: That's the problem with 'evidence' predicting earthquakes and volcanic eruptions — it's nowhere near 100% reliable. There may be shed-loads of people living near a volcano or a fault line — it'd be impossible to evacuate them all every time scientists thought there might possibly be an eruption or an earthquake some time soon — it just wouldn't work.

Evidence for Plate Tectonics

Q1 Choose from the words below to complete the paragraph.

continents Sangria evolution living creatures fast-food restaurants plate tectonics fossils Pangaea

All the continents were once joined in a huge land mass called ....
It is now believed that this land mass slowly split up to form the ....
This is the theory of .... It explains why continents thousands of miles apart have identical .... and
....

Q2 Fossils of the prehistoric fern **Glossopteris** have been found in India, Australia, Africa, South America and Antarctica.

a) Suggest why this distribution cannot be explained by seed dispersal.

....

India
Africa
Australia
Antarctica

b) What would be a better explanation of this?

....

....

Q3 The diagram shows **rock sequences** from four different continents.

Continent A　Continent B　Continent C　Continent D

a) Which two continents do you think were joined at one time? Explain your answer.

....

b) Suggest what other evidence you could look for in the rocks to back this up.

....

c) Describe two other pieces of evidence (apart from that found in rocks) which support the theory that the continents were once all joined.

1. ....

....

2. ....

....

The Three Different Types of Rock

Q1 Join up each **rock type** with the correct **method of formation** and an **example**.

ROCK TYPE	METHOD OF FORMATION	EXAMPLE
igneous rocks	formed from layers of sediment	granite
metamorphic rocks	formed when magma cools	limestone
sedimentary rocks	formed under intense heat and pressure	marble

Q2 Circle the correct words to complete the passage below.

Metamorphic / **Igneous** rock is formed when magma pushes up into (or through) the **crust** / **mantle** and cools.

If the magma cools before it reaches the surface it will cool **slowly** / **quickly**, forming **big** / **small** crystals. This rock is known as **extrusive** / **intrusive** igneous rock. Examples of this are **basalt** / **granite** and **gabbro** / **rhyolite**.

However the magma that reaches the surface will cool **slowly** / **quickly**, forming **big** / **small** crystals. This rock is known as **extrusive** / **intrusive** igneous rock. Examples of this are **basalt** / **granite** and **gabbro** / **rhyolite**.

Q3 Erica notices that the stonework of her local church contains tiny fragments of **sea shells**.

a) Suggest an explanation for this.

....

....

b) Describe how sedimentary rock is 'cemented' together.

....

....

c) Powdered limestone and powdered marble react with other chemicals, such as hydrochloric acid, in an identical fashion. Explain this.

....

Top Tips: You might think that rocks are just boring lumps of.... rock. But you'd be wrong — rocks are actually boring lumps of different kinds of rock. And the kind of rock they are depends on how they're formed — and this is the stuff you need to make sure you know.

The Evolution of the Atmosphere

Q1 Tick the boxes next to the sentences below that are **true**.

When the Earth was formed its surface was molten. ☐

The Earth's early atmosphere is thought to have been mostly oxygen. ☐

When oxygen started building up in the atmosphere, all organisms began to thrive. ☐

When some plants died and were buried under layers of sediment, the carbon they had removed from the atmosphere was locked up as fossil fuels. ☐

The development of the ozone layer meant the Earth's temperature reached a suitable level for complex organisms like us to evolve. ☐

Q2 Draw lines to put the statements in the **right order** on the time line. One is done for you.

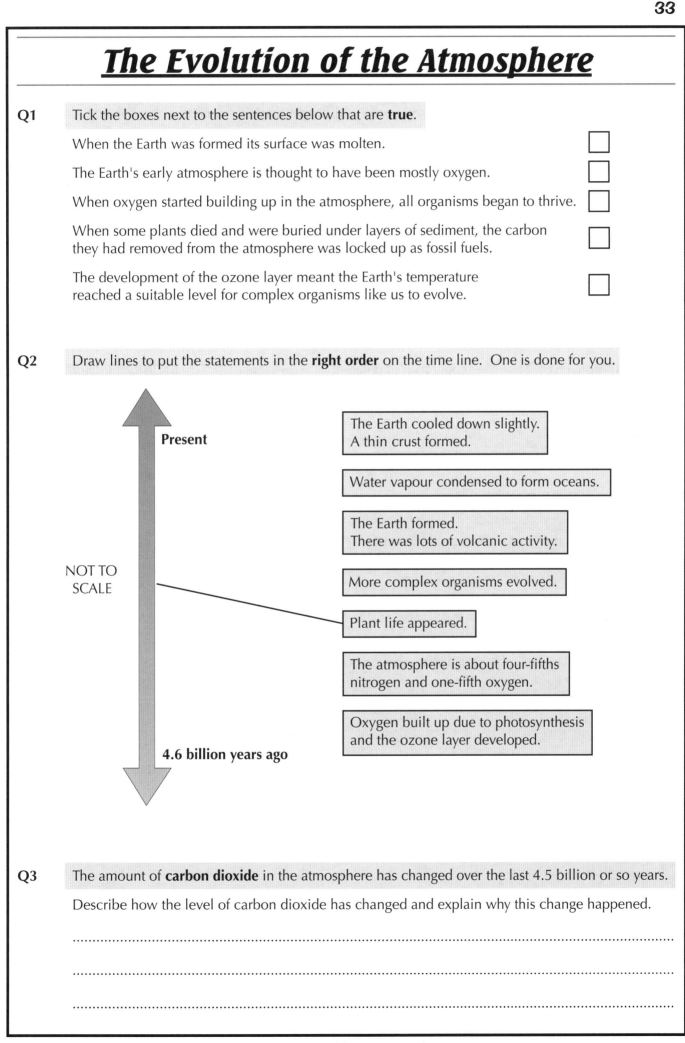

Present

NOT TO SCALE

4.6 billion years ago

The Earth cooled down slightly. A thin crust formed.

Water vapour condensed to form oceans.

The Earth formed. There was lots of volcanic activity.

More complex organisms evolved.

Plant life appeared.

The atmosphere is about four-fifths nitrogen and one-fifth oxygen.

Oxygen built up due to photosynthesis and the ozone layer developed.

Q3 The amount of **carbon dioxide** in the atmosphere has changed over the last 4.5 billion or so years.

Describe how the level of carbon dioxide has changed and explain why this change happened.

....

....

....

The Evolution of the Atmosphere

Q4 The pie chart below shows the proportions of **different gases** in the Earth's atmosphere today.

a) Add the labels '**Nitrogen**', '**Oxygen**', and '**Carbon dioxide and other gases**'.

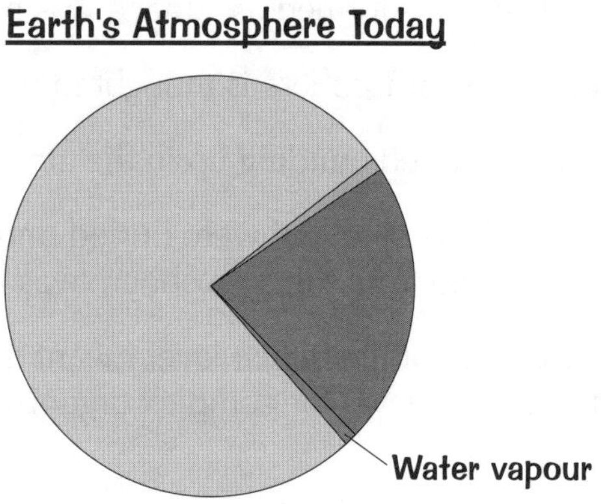

Earth's Atmosphere Today

Water vapour

b) Give the approximate percentages of the following gases in the air today:

Nitrogen: ....%

Oxygen: ....%

c) This pie chart shows the proportions of different gases that it's thought were in the Earth's atmosphere 4.5 billion years ago.

Earth's Atmosphere 4.5 Billion Years Ago

Carbon dioxide

Nitrogen

Other gases

Water vapour

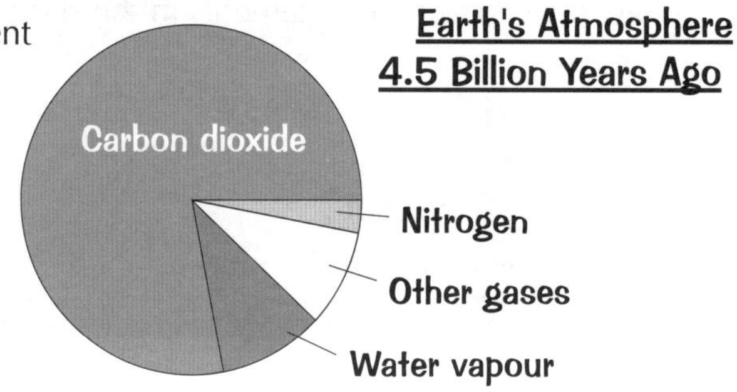

Describe the main differences between today's atmosphere and the atmosphere 4.5 billion years ago.

....

....

d) Explain why the amount of water vapour has decreased.

....

....

What did the water vapour change into?

e) Explain how oxygen was introduced into the atmosphere.

....

f) Describe two effects of the rising oxygen levels in the atmosphere.

1. ....

....

2. ....

....

Atmospheric Change

Q1 There is a scientific theory that says that the water on Earth came from **comets**, not volcanoes.

Why is this theory not accepted by many scientists?

....

....

Q2 The graphs below show the changes in atmospheric **carbon dioxide** levels and global **temperature** since 1850.

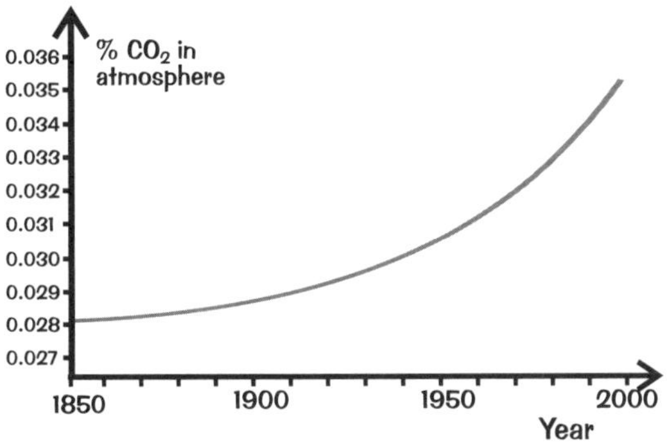

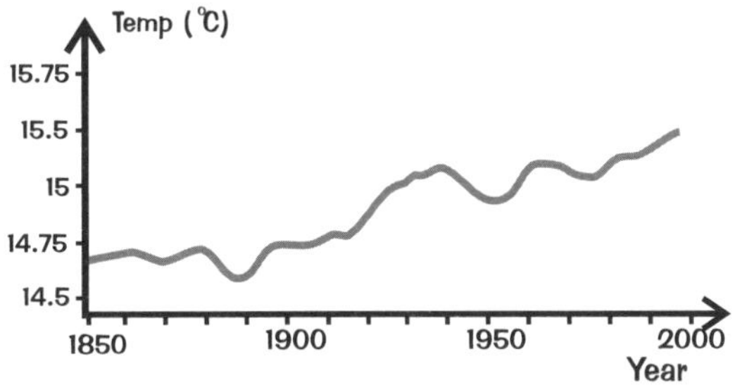

% CO_2 in atmosphere: 0.027, 0.028, 0.029, 0.030, 0.031, 0.032, 0.033, 0.034, 0.035, 0.036; Year: 1850, 1900, 1950, 2000

Temp (°C): 14.5, 14.75, 15, 15.5, 15.75; Year: 1850, 1900, 1950, 2000

a) State **two** human activities that have contributed to the rise in carbon dioxide levels over the last 150 years.

....

b) Compare the carbon dioxide and temperature graphs. What do they tell you about the **relationship** between CO_2 levels and temperature?

....

Q3 Evidence shows that the **ozone layer** is **thinning** and in places holes have developed. This has been linked to an increase in **skin cancer** over the past thirty years.

a) Explain how the thinning of the ozone layer may have contributed to the rise in skin cancer.

....

....

b) Do these facts prove that the thinning of the ozone layer has caused the rise in skin cancer cases? Explain your answer.

....

....

The Carbon Cycle and Climate Change

Q1 Underline the statements below about the greenhouse effect that are **true**.

The greenhouse effect is needed for life on Earth as we know it.

Greenhouse gases include carbon dioxide and sulfur dioxide.

The greenhouse effect causes acid rain.

Increasing amounts of greenhouse gases lead to global warming.

Q2 Here is a diagram of the **carbon cycle**.

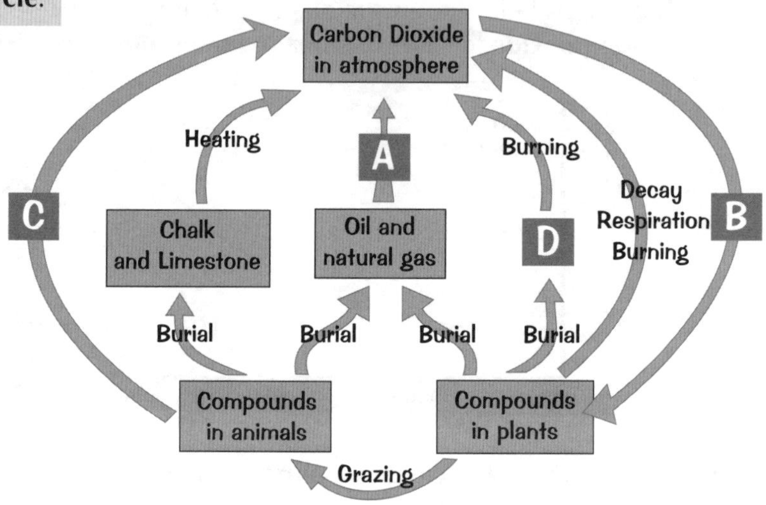

a) What is process A?

....

b) What is process B?

....

c) Process C could be decay.
What else could it be?

....

d) What is substance D?

....

Q3 The Earth receives energy from the **Sun**. It radiates much of this energy back out to space.

a) Explain the role of the greenhouse gases in keeping the Earth warm.

....

....

b) In recent years the amounts of greenhouse gases in the atmosphere have increased.
Explain how this leads to global warming.

....

....

Q4 **Carbon** that was once part of Henry VIII could now be part of you.

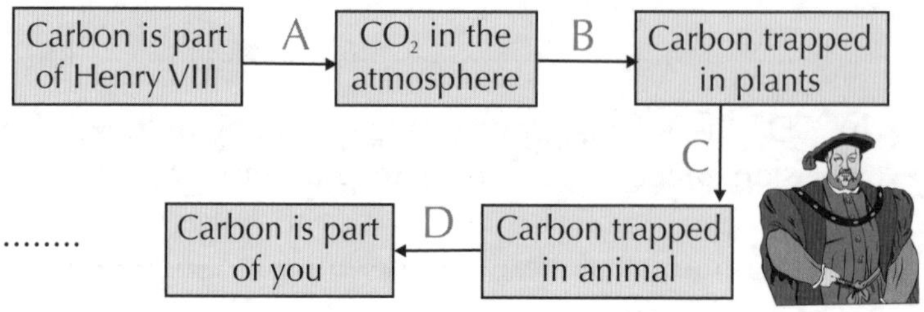

Name each process in the diagram.

A ....

B ....

C ....

D ....

Human Impact on the Environment

Q1 Circle the correct word to complete each sentence below.

a) The human population now is **bigger** / **smaller** than it was 1000 years ago.

b) The growth of the human population now is **slower** / **faster** than it was 1000 years ago.

c) The human impact on the environment now is **less** / **greater** than it was 1000 years ago.

Q2 **Waste** produced by humans can pollute water, land and air.

a) Give **two** ways that each of the following can be polluted by human activities.

i) Water: ....

....

ii) Land: ....

....

iii) Air: ....

....

b) Other than dumping waste, name two human activities that reduce the amount of land available for other animals.

1. .... 2. ....

Q3 One way to assess a person's impact on the Earth is to use an **ecological footprint**. This involves calculating **how many Earths** would be needed if everyone lived like that person. It takes into account things like the amount of **waste** the person produces and how much **energy** they use.

a) Two men calculate their ecological footprints. Eight Earths would be needed to support everyone in the way John lives. Half an Earth would be enough to support everyone in the way Derek lives.

i) One of the men lives in a UK city, and one in rural Kenya. Who is more likely to live where?

....

ii) Tick any of the following that are possible reasons for the difference in results.

- [ ] **John buys more belongings, which use more raw materials to manufacture.**
- [ ] **John has central heating in his home and Derek has a wood fire.**
- [ ] **John throws away less waste.**
- [ ] **John drives a car and Derek rides a bicycle.**

b) Suggest one thing John could do to reduce the size of his ecological footprint.

....

Air Pollution and Acid Rain

Q1 Choose from the words and phrases below to complete the paragraph.

nitric **sulfur dioxide** **sulfuric** **nitrogen oxides** **acid rain**

When fossil fuels are burnt the gas .... is produced from impurities in the fuel. When it combines with moisture in the air, ... acid is produced — this falls as acid rain.

In the high temperature inside a car engine, nitrogen and oxygen from the air react together to produce .... These react with moisture to produce .... acid, which is another cause of acid rain.

Q2 **Acid rain** causes a variety of problems.

I definitely felt a spot of rain then.

a) Why might architects choose **not** to build from limestone?

....

b) Give **two** other consequences of acid rain.

....

c) What can be done in a fossil-fuel **power station** to reduce the contribution it makes to acid rain?

....

Q3 **Exhaust fumes** from cars and lorries often contain **carbon monoxide** (CO).

a) Why is CO more likely to be formed in engines than if the same fuel were burnt in the open air?

....

b) Why is carbon monoxide so dangerous?

....

Q4 **Catalytic converters** on motor vehicles reduce the amount of harmful gases they release.

a) Complete this word equation to show the reaction that occurs in a catalytic converter:
carbon monoxide + nitrogen monoxide → .... + ...

b) Write a balanced symbol equation for this reaction. ....

c) What catalyst is used for this reaction? ....

Top Tips: The best way to prevent acid rain damage is to reduce the amount of sulfur dioxide that we release into the atmosphere. When acid rain does fall there are some ways of reducing the amount of damage it causes, such as adding powdered limestone to affected lakes.

Protecting the Atmosphere

Q1 Biogas is a mixture of methane (CH_4) and carbon dioxide. It is produced by microorganisms digesting waste material.

a) What products are formed when **biogas** is burnt?

....

b) What would be the main problem with using a **biogas generator** in Iceland?

....

Q2 Isobella is trying to decide which hydrocarbon, A or B, is the best one to use as a fuel. She tests the **energy content** of the hydrocarbons by using them to heat 50 cm^3 of water from 25 °C to 40 °C. The results of this experiment are shown in the table.

Hydrocarbon	Initial Mass (g)	Final Mass (g)	Mass of Fuel Burnt (g)
A	98	92	
B	102	89	

a) Complete the table by calculating the mass of fuel that was burned in each case.

b) Which fuel contains more energy per gram? ....

c) Name four other things Isobella should consider when choosing the best fuel to use.

....

....

Q3 Scientists try to **predict** what the Earth's **temperature** will be in the future. There are various steps involved in doing this.

a) Describe briefly how monitoring stations and computers are involved in making these predictions.

....

....

b) Explain why these predictions may not be accurate.

....

....

c) Suggest how the predictions can be tested.

....

....

Protecting the Atmosphere

Q4 In response to scientific advice about climate change, some governments have adopted the '**precautionary principle**'.

a) Describe what is meant by the term 'precautionary principle'.

....

....

b) Apply the precautionary principle to combating climate change.

....

....

c) Give two ways that the amount of greenhouse gas added to the atmosphere could be reduced.

1. ....

2. ....

Q5 **Hydrogen** is often talked about as the 'fuel of the future'.

a) What waste product is produced when hydrogen is burned?

....

b) Why is it better for the environment if hydrogen is burned rather than petrol?

....

....

c) Currently, most of the vehicles that can use hydrogen as a fuel are demonstration vehicles that are being developed by scientists.

Explain the problems that will have to be overcome before the public will be able to use hydrogen-powered vehicles on a large scale.

....

....

....

Think about storage of hydrogen and the costs involved.

Top Tips: Scientists are looking at the ways people are damaging the environment and trying to work out how to reduce the damage. But even when they've agreed on the best way to fix a problem, governments often don't like it — because it's **expensive** or **unpopular** in the short term.

Recycling Materials

Q1 Explain what is meant by **sustainable development**.

....

....

Q2 Give three advantages of recycling materials like **glass** and **plastics**.

1. ....
2. ....
3. ....

Q3 Why should we **recycle paper** when new trees can be grown to replace those cut down?

....

Q4 Below is some information about **aluminium**, a widely used metal.

- **For every 4 kg of bauxite (aluminium ore) mined, 1 kg of aluminium is produced.**
- **Bauxite mines are often located in rainforests.**
- **Extracting aluminium from bauxite requires huge quantities of electricity.**
- **An aluminium can weighs about 20 g.**

a) **i)** How much ore has to be mined to produce 1 tonne (1000 kg) of aluminium? ....

ii) Australians used over 3 billion aluminium cans in 2002. How much aluminium is this in tonnes?

....

iii) How many tonnes of bauxite were mined to supply Australians with aluminium cans in 2002?

....

b) Briefly describe the environmental consequences of:

i) mining bauxite ....

....

ii) extracting the aluminium ....

....

iii) **not** recycling the cans ....

....

The Periodic Table and Electron Shells

Q1 Select from these **elements** to answer the following questions.

iodine nickel silicon sodium radon krypton calcium

a) Which two elements are in the same group? .... and ....

b) Name two elements which are in Period 3. .... and ....

c) Name an alkali metal. ....

d) Name a transition metal. ....

e) Name an element with seven electrons in its outer shell. ....

f) Name a non-metal which is not in Group 0. ....

Q2 Tick the correct boxes to show whether these statements are **true** or **false**.

		True	False
a)	Elements in the same **group** have the same number of electrons in their outer shell.	☐	☐
b)	The periodic table shows the elements in order of ascending **atomic mass**.	☐	☐
c)	Each **column** in the periodic table contains elements with similar properties.	☐	☐
d)	The periodic table is made up of all the known compounds.	☐	☐
e)	Each new period in the periodic table represents another full shell of electrons.	☐	☐

Q3 Elements in the same group undergo **similar reactions**.

a) Tick the pairs of elements that would undergo similar reactions.

A potassium and rubidium ☐ **C** calcium and oxygen ☐

B helium and fluorine ☐ **D** nitrogen and arsenic ☐

b) Explain why fluorine and chlorine undergo similar reactions.

....

....

Q4 Complete the following table.

	Alternative Name	Number of Electrons in Outer Shell
Group I		
Group VII		
Group 0		*

* excluding helium

Electron Shells

Q1 a) Tick the boxes to show whether the statements are **true** or **false**.

	True	False
i) Electrons occupy shells.	☐	☐
ii) The highest energy levels are always filled first.	☐	☐
iii) Atoms are most stable when they have partially filled shells.	☐	☐
iv) Noble gases have full outer shells of electrons.	☐	☐
v) Reactive elements have full outer shells of electrons.	☐	☐

b) Write out corrected versions of the **false** statements.

....

....

....

Q2 Identify **two** things that are wrong with this diagram.

1. ....

....

2. ....

....

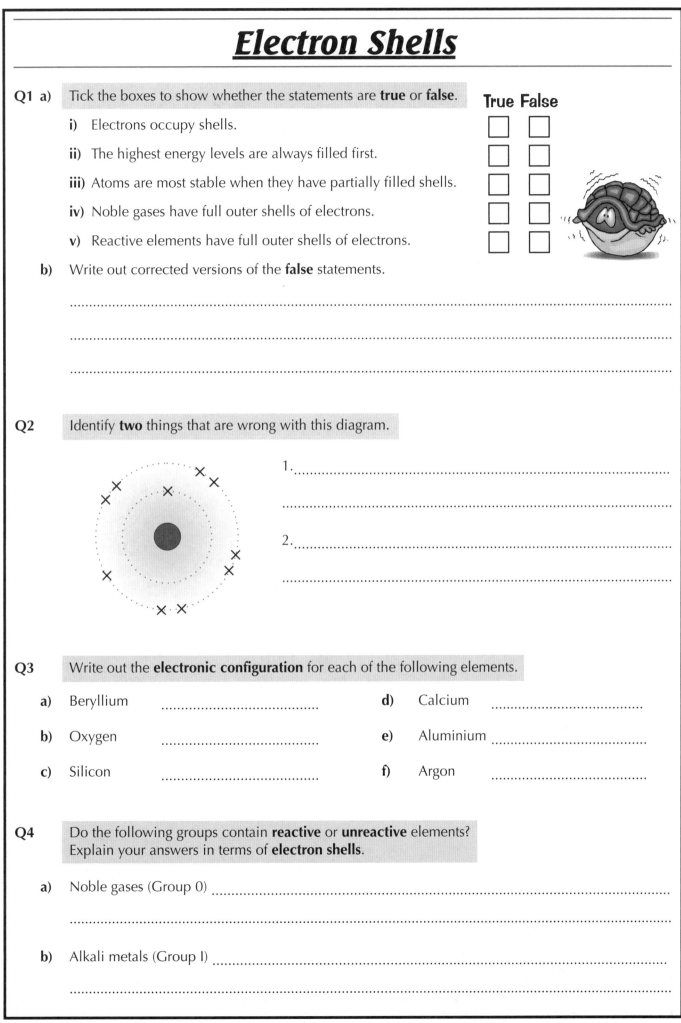

Q3 Write out the **electronic configuration** for each of the following elements.

a) Beryllium ....

b) Oxygen ....

c) Silicon ....

d) Calcium ....

e) Aluminium ....

f) Argon ....

Q4 Do the following groups contain **reactive** or **unreactive** elements? Explain your answers in terms of **electron shells**.

a) Noble gases (Group 0) ....

....

b) Alkali metals (Group I) ....

....

Electron Shells

Q5 **Chlorine** has an atomic number of 17.

a) What is chlorine's electron configuration? ....

b) Draw the electrons on the shells in the diagram. 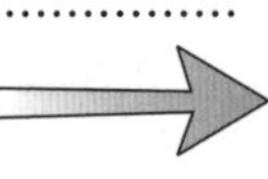

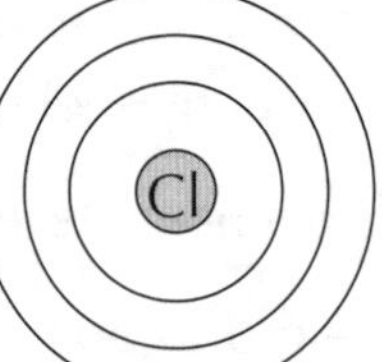

c) Why does chlorine react readily?

....

Q6 Draw the **full electronic arrangements** for these elements. (The first three have been done for you.)

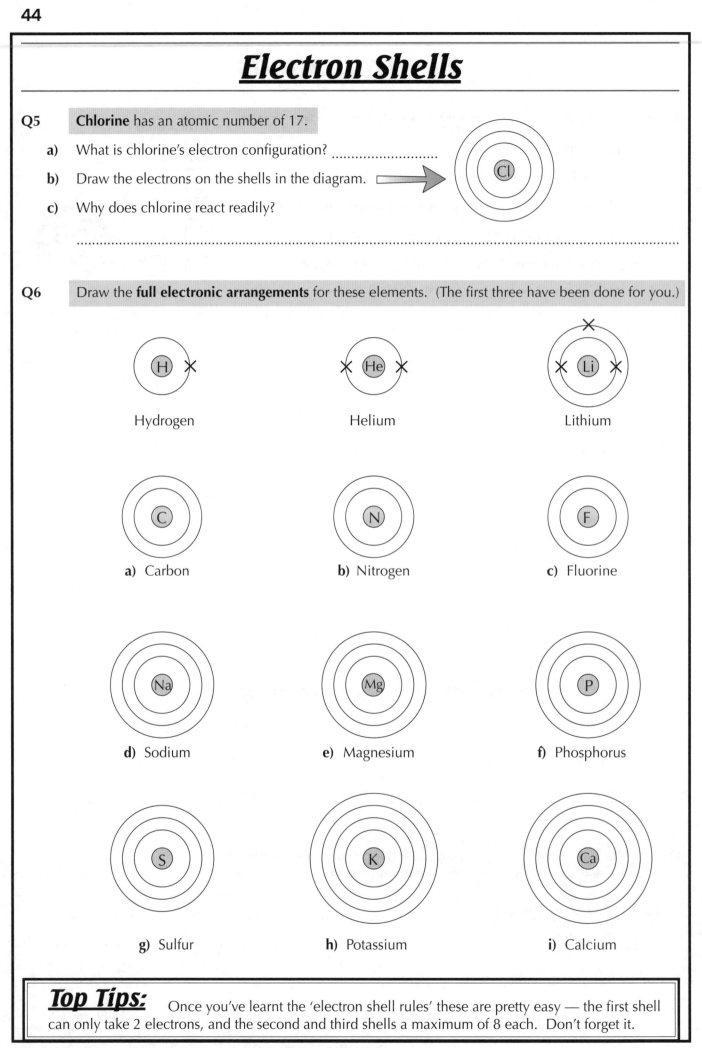

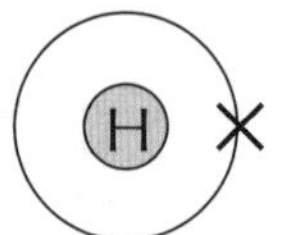

Hydrogen

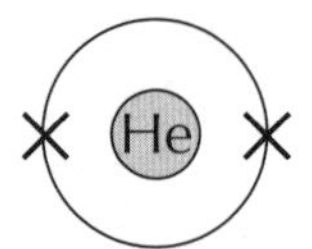

Helium

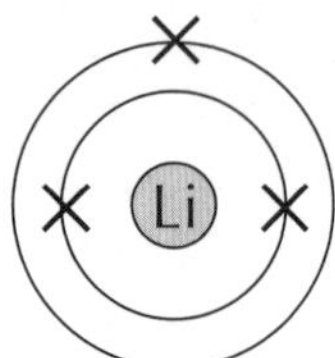

Lithium

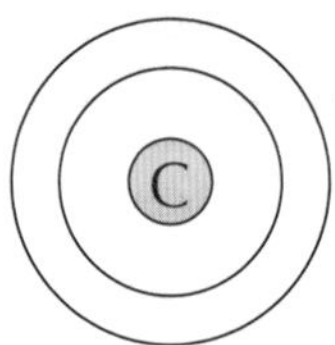

a) Carbon

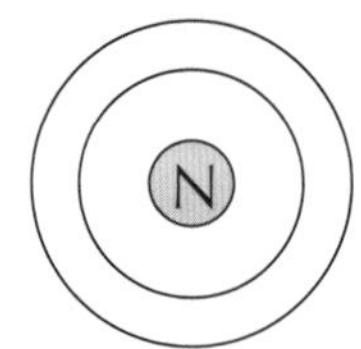

b) Nitrogen

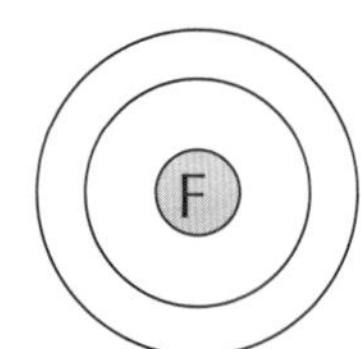

c) Fluorine

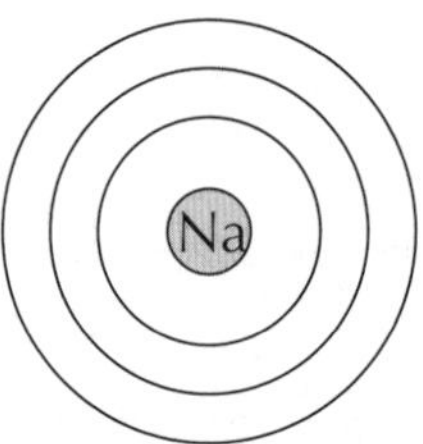

d) Sodium

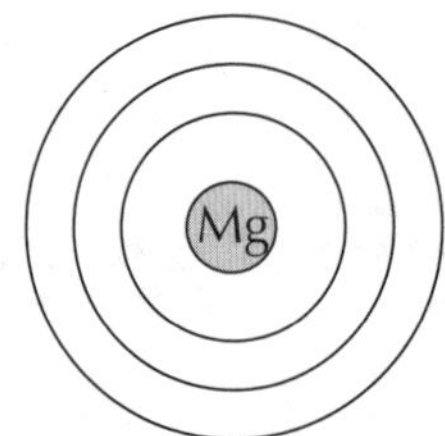

e) Magnesium

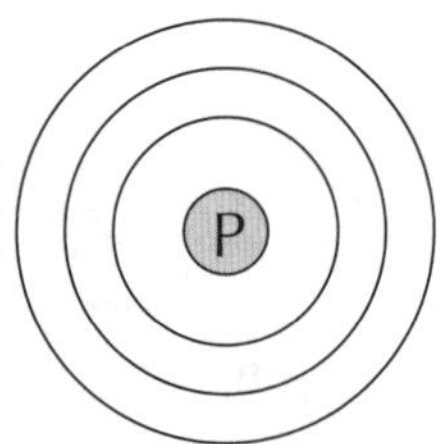

f) Phosphorus

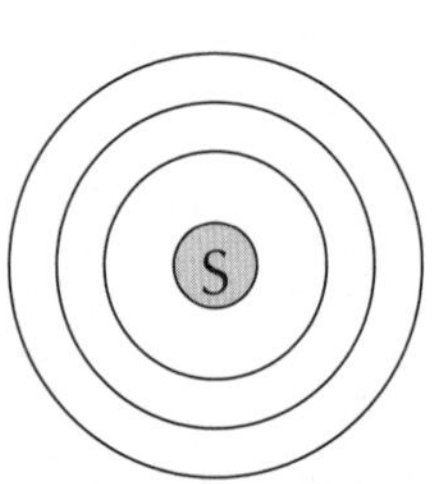

g) Sulfur

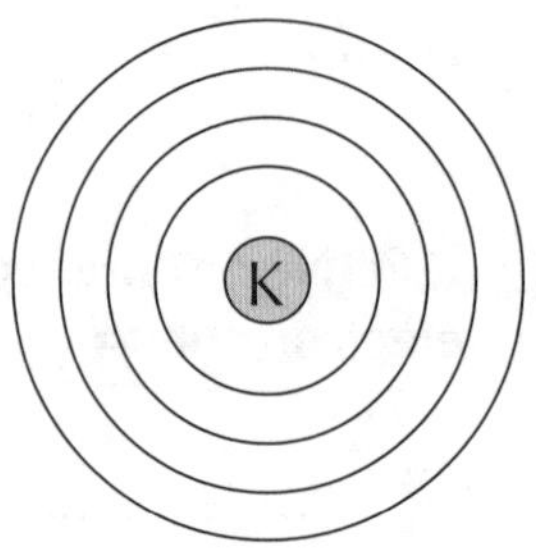

h) Potassium

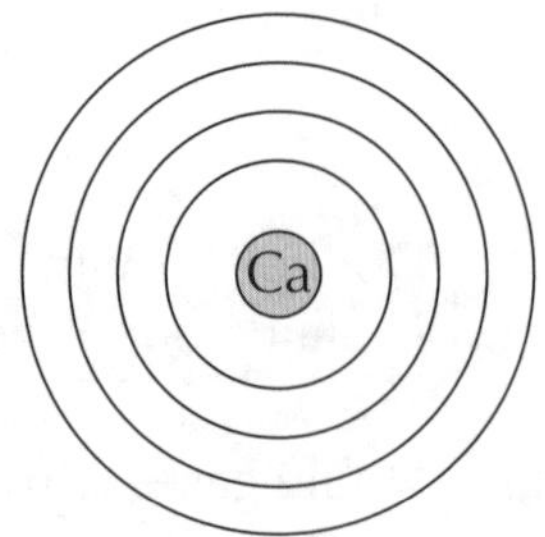

i) Calcium

Top Tips: Once you've learnt the 'electron shell rules' these are pretty easy — the first shell can only take 2 electrons, and the second and third shells a maximum of 8 each. Don't forget it.

Ionic Bonding

Q1 Fill in the gaps in the sentences below by choosing the correct words from the box.

protons	charged particles	repelled by	
electrons	ions	attracted to	neutral particles

a) In ionic bonding atoms lose or gain .... to form ....

b) Ions are ....

c) Ions with opposite charges are strongly .... each other.

Q2 Tick the correct boxes to show whether the following statements are **true** or **false**.

		True	False
a)	Metals generally have lower numbers of electrons in their outer shells than non-metals.	☐	☐
b)	Metals form negatively charged ions.	☐	☐
c)	Elements in Group VII gain electrons when they react.	☐	☐
d)	Atoms form ions because they are more stable when they have full outer shells.	☐	☐
e)	Elements in Group 0 form ions very readily.	☐	☐

Q3 Use this **diagram** to answer the following questions.

Na Cl + − Na Cl

a) Which group of the periodic table does **sodium** belong to? ....

b) How many electrons does **chlorine** need to gain to get a full outer shell of electrons? ....

c) What is the charge on a **sodium ion**? ....

d) What is the chemical formula of **sodium chloride**? ....

Q4 Here are some **elements** and the **ions** they form:

Make sure the charges on the ions balance.

beryllium, Be^{2+} **potassium, K^+** **iodine, I^-** **sulfur, S^{2-}**

Write down the formulas of four compounds which can be made using these elements.

1. .... 2. ....

3. .... 4. ....

Ionic Bonding

Q5 Magnesium and oxygen react to form **magnesium oxide**, an **ionic** compound.

a) Draw a 'dot and cross' diagram showing the formation of magnesium oxide from magnesium and oxygen atoms.

b) What name is given to the structure of magnesium oxide?

....

c) Circle the correct words to explain why magnesium oxide has a high melting point.

Magnesium oxide has very **strong** / **weak** chemical bonds between the **negative** / **positive** magnesium ions and the **negative** / **positive** oxygen ions. This means that it takes a **small** / **large** amount of energy to break the bonds and melt the compound.

Q6 Mike conducts an experiment to find out if **sodium chloride** conducts electricity. He tests the compound when it's solid, when it's dissolved in water and when it's molten.

a) Complete the table of results opposite.

b) Explain your answers to part **a)**.

	Conducts electricity?
When solid	
When dissolved in water	
When molten	

....

....

....

Q7 Draw '**dot and cross**' diagrams showing the formation of the following ionic compounds:

a) sodium oxide

b) magnesium chloride

Ions and Formulas

Q1 Complete the following sentences by circling the correct words.

a) Elements in Group I **lose** / **gain** electrons to form ions.

b) Elements in Group VII **lose** / **gain** electrons to form ions.

c) Positive ions are called **anions** / **cations**.

Q2 Atoms can **gain** or **lose** electrons to get a full outer shell.

a) How many **electrons** do the following elements need to **lose** in order to get a **full outer shell**?

i) Lithium .... **ii)** Calcium .... **iii)** Potassium ....

b) How many **electrons** do the following elements need to **gain** in order to get a **full outer shell**?

i) Oxygen .... **ii)** Chlorine .... **iii)** Fluorine ....

Q3 Write the **electron configurations** for the following ions and draw the **electrons** on the shells. (The first one's been done for you.)

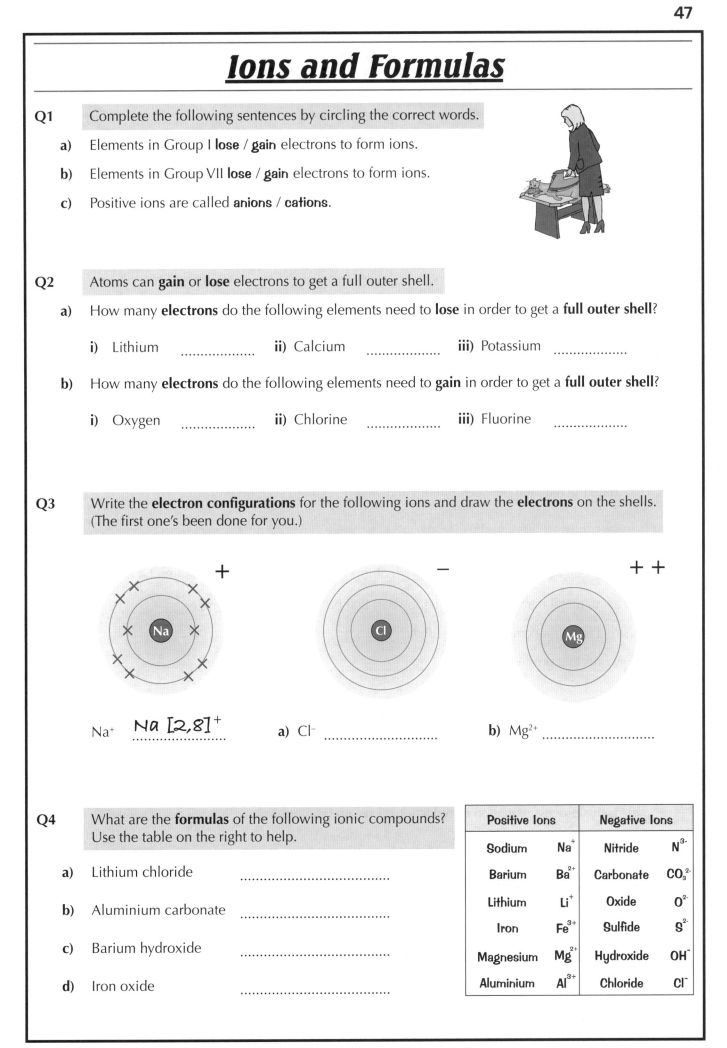

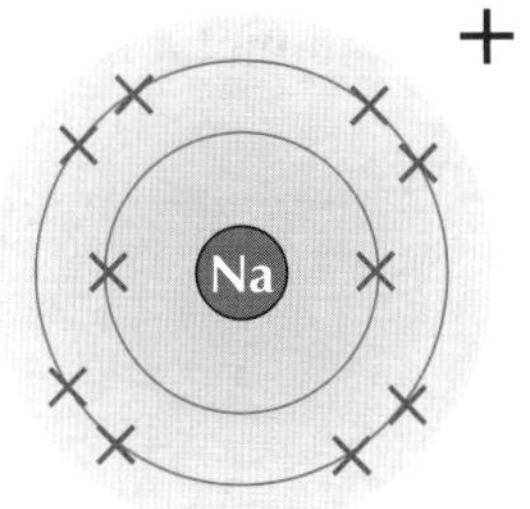

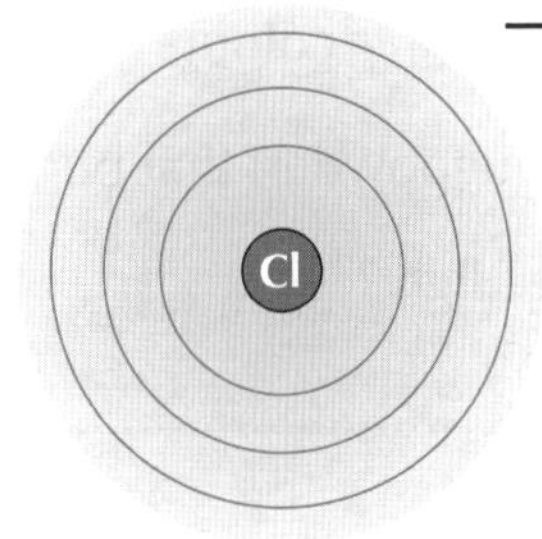

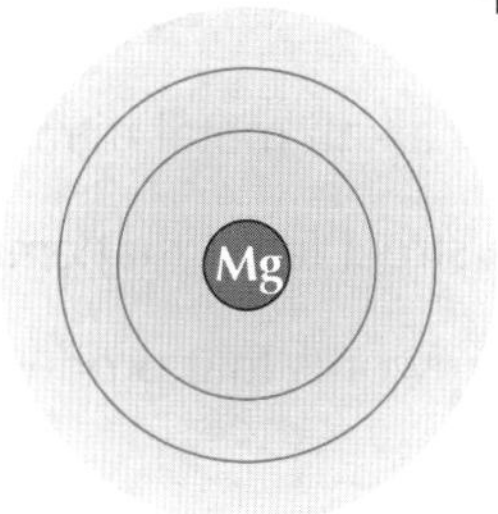

Na^+ 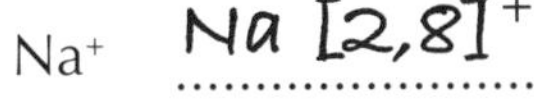

a) Cl^- ....

b) Mg^{2+} ....

Q4 What are the **formulas** of the following ionic compounds? Use the table on the right to help.

a) Lithium chloride ....

b) Aluminium carbonate ....

c) Barium hydroxide ....

d) Iron oxide ....

Positive Ions		Negative Ions	
Sodium	Na^+	Nitride	N^{3-}
Barium	Ba^{2+}	Carbonate	CO_3^{2-}
Lithium	Li^+	Oxide	O^{2-}
Iron	Fe^{3+}	Sulfide	S^{2-}
Magnesium	Mg^{2+}	Hydroxide	OH^-
Aluminium	Al^{3+}	Chloride	Cl^-

Group 1 — Alkali Metals

Q1 Archibald put a piece of **lithium** into a beaker of water.

a) Explain why the lithium floated on top of the water.

b) After the reaction had finished, Archibald tested the water with universal indicator. What colour change would he see, and why?

"squeaky pop!"

Archibald

c) Write a **balanced symbol equation** for the reaction between lithium and water.

d) i) Write a word equation for the reaction beetween rubidium and water.

ii) Would you expect the reaction between rubidium and water to be **more** or **less** vigorous than the reaction between lithium and water? Explain your answer.

Q2 Alkali metal compounds emit particular **colours** when heated. Which **alkali metal** is present in:

a) an alkali metal nitrate (found in gunpowder) that burns with a lilac flame?

b) a street lamp that emits an orange light?

c) fireworks that produce red flames?

Q3 Sodium and potassium are both very **reactive**.

a) i) Write a balanced symbol equation to show the formation of a sodium ion from a sodium atom.

ii) Is this process oxidation or reduction? Explain your answer.

b) Why do sodium and potassium have similar properties?

c) Why is potassium more reactive than sodium?

Group VII — Halogens

Q1 Chlorine and bromine are both **halogens**.

a) **i)** Write a balanced symbol equation to show the formation of chloride ions from a Cl_2 molecule.

....

ii) Is this process oxidation or reduction? Explain your answer.

....

b) Why is bromine less reactive than chlorine?

....

....

Q2 Equal volumes of **bromine water** were added to two test tubes, each containing a different **potassium halide solution**. The results are shown in the table.

SOLUTION	RESULT
potassium chloride	no colour change
potassium iodide	colour change

a) Explain these results.

....

....

b) Write a **balanced symbol equation** for the reaction in the potassium iodide solution.

....

c) Would you expect a reaction between:

i) bromine water and potassium astatide? ....

ii) bromine water and potassium fluoride? ....

Q3 **Sodium** was reacted with **bromine vapour** using the equipment shown. White crystals of a new solid were formed during the reaction.

a) Name the white crystals.

....

b) Write a balanced symbol equation for the reaction.

....

c) Would you expect the reaction between sodium and bromine vapour to be faster or slower than a similar reaction between sodium and iodine vapour? Explain your answer.

....

Covalent Bonding

Q1 Indicate whether each statement is **true** or **false**.

		True	False
a)	Covalent bonding involves sharing electrons.	☐	☐
b)	Atoms react to gain a full outer shell of electrons.	☐	☐
c)	Some atoms can make both ionic and covalent bonds.	☐	☐
d)	Hydrogen can form two covalent bonds.	☐	☐
e)	Carbon can form four covalent bonds.	☐	☐

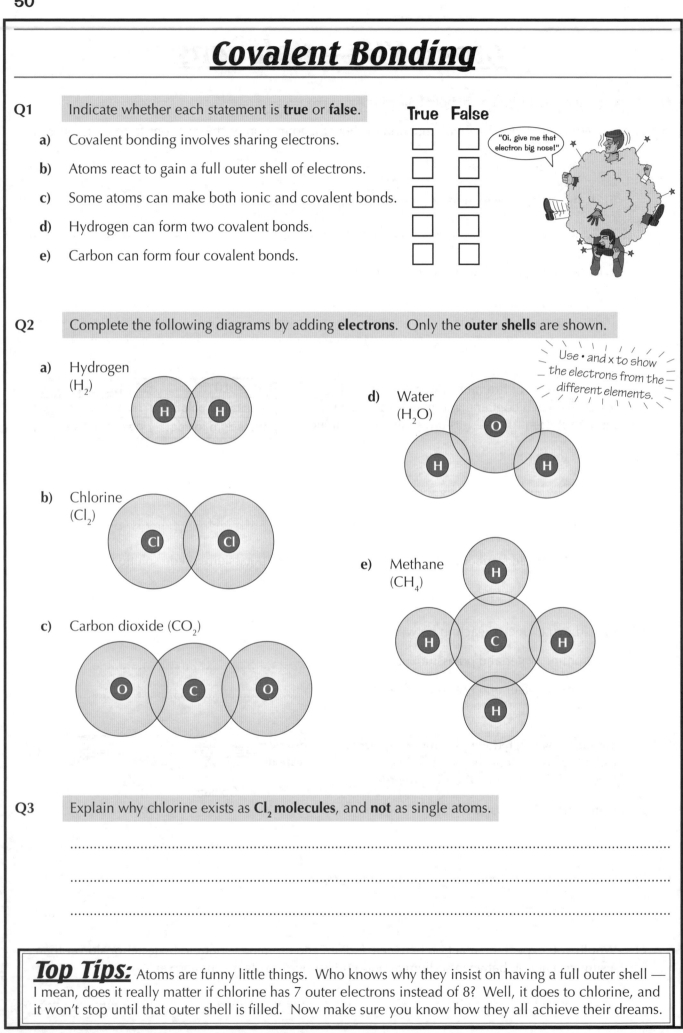

"Oi, give me that electron big nose!"

Q2 Complete the following diagrams by adding **electrons**. Only the **outer shells** are shown.

Use • and x to show the electrons from the different elements.

a) Hydrogen (H_2)

H H

b) Chlorine (Cl_2)

Cl Cl

c) Carbon dioxide (CO_2)

O C O

d) Water (H_2O)

O H H

e) Methane (CH_4)

H H C H H

Q3 Explain why chlorine exists as **Cl_2 molecules**, and **not** as single atoms.

....

....

....

Top Tips: Atoms are funny little things. Who knows why they insist on having a full outer shell — I mean, does it really matter if chlorine has 7 outer electrons instead of 8? Well, it does to chlorine, and it won't stop until that outer shell is filled. Now make sure you know how they all achieve their dreams.

Giant Covalent Structures

Q1 Circle the correct words to complete the following paragraph.

> Giant covalent structures contain **charged ions / uncharged atoms**. The covalent bonds between the atoms are **strong / weak**. Giant covalent structures have **high / low** melting points and they are usually **soluble / insoluble** in water.

Q2 **Graphite** and **diamond** are both made entirely from **carbon**, but have different properties.

a) Explain why graphite is a good conductor of electricity.

....

....

....

b) Explain how diamond's structure makes it hard.

....

....

....

Q3 The different **forms** of carbon have different **properties** and **uses**. Match each of the two uses in pink to one of the forms of carbon listed below. State the property that justifies your choice.

glass-cutting tool **pencils**

a) **Graphite** Use: ....

Property: ....

b) **Diamond** Use: ....

Property: ....

Q4 Grains of **sand** are giant covalent structures.

a) What is the chemical name for sand? ....

b) Which two elements does it contain? .... and ....

c) Sand can be melted with limestone and sodium carbonate to make a useful product. What is it?

....

Simple Molecular Covalent Structures

Q1 Fill in the blanks in the following paragraph by choosing words from the list.

weak **hard** **small** **easy** **large** **strong**

Simple molecular substances are made from .... molecules. The covalent bonds that hold the atoms together are .... but the forces between the molecules are quite .... Because of this it is fairly .... to separate the molecules.

Q2 Hydrogen and chlorine share electrons to form a molecule called **hydrogen chloride**.

Predict two properties that hydrogen chloride will have.

1. ....

2. ....

Q3 Complete the following sentences by circling the correct option, and explain each property.

a) The melting and boiling points of simple molecular substances are **low** / **high**.

....

b) Simple molecular substances **conduct** / **don't conduct** electricity.

....

Q4 The table below shows the **atomic numbers** and **melting points** for three **halogens**.

Halogen	Atomic no.	Melting pt.
Fluorine	9	– 220 °C
Bromine	35	–7 °C
Iodine	53	114 °C

Melting point (°C): 120, 100, 80, 60, 40, 20, 0, –20, –40, –60, –80, –100, –120, –140, –160, –180, –200, –220

Atomic number: 10, 20, 30, 40, 50

a) Plot the data on the axes given and draw a line of best fit.

b) Describe and explain the relationship between atomic number and melting point for the halogens.

....

....

....

....

Group 0 — Noble Gases

Q1 Where are the noble gases located in the **periodic table**?

....

Q2 Circle the correct word(s) in each pair to complete the following sentences about **noble gases**.

a) The elements of Group 0 are all **brightly coloured** / **colourless**.

b) The Group 0 elements all have **high** / **low** boiling points.

c) All Group 0 elements need to **gain** / **lose** / **neither gain nor lose** electrons to fill their outer shells.

Q3 The noble gases were **discovered** long after many of the other elements.

a) Why did it take scientists so long to discover the noble gases?

....

b) Explain why the noble gases are unreactive. ....

....

c) What can be done to noble gases to make them visible?

....

Q4 Match up the names of the noble gases to their correct **uses** below.

Neon — **used in electric light bulbs**

Helium — **used in airships and balloons**

Argon — **used in signs**

Q5 Three gases are tested to try to **identify** them. The results of the tests are shown in the table below.

TEST	TEST RESULTS		
	GAS A	GAS B	GAS C
Balloon filled with the gas is released	Balloon sinks	Balloon rises	Balloon sinks
Lighted match dropped into jar of the gas	Goes out	Goes out	Burns brightly
Electric current passed through gas	Bright green-white glow	Bright yellow glow	Faint blue-green glow

a) One of gases A, B and C is helium. Which one is it? Explain your answer.

....

b) One of the gases is **not** a member of Group 0 of the periodic table.
Which of the three gases do you think this is? Explain your answer.

....

Metallic Structures

Q1 Complete the following sentences by choosing from the words in the box.

hammered	conductors	low	high	insulators	malleable	folded

a) Metals have a .... tensile strength.

b) Metals are usually excellent ....

c) Metals can be .... into different shapes because they are ....

Q2 Draw lines to match the transition metal to the process it **catalyses**.

iron — converting natural oils into fats

nickel — ammonia production

Q3 Under normal conditions **all** metals have **electrical resistance**.

a) Describe how electrical resistance causes energy to be wasted.

....

....

b) What is a superconductor? How are they made?

....

....

c) Give three possible uses of superconducting wires.

1. ....

2. ....

3. ....

d) Explain a drawback of using today's superconductors.

....

....

Top Tips: The properties of metals are down to the structure they all have in common — a giant structure held together with metallic bonds and a sea of free electrons. That's not to say all metals have exactly the same properties though. You can cut alkali metals with a knife — wouldn't work on iron.

Nanomaterials

Q1 A molecule of **buckminsterfullerene** is made up of 60 carbon atoms.

a) What is the **molecular formula** of buckminsterfullerene? ....

b) How many covalent bonds does each carbon atom form? ....

c) Can buckminsterfullerene conduct electricity? Explain your answer.

....

....

Q2 **Nanoparticles** have the potential to be extremely **useful** materials. Describe one use of:

a) zinc oxide nanoparticles ....

b) silver nanoparticles ....

c) carbon nanoparticles ....

Q3 **Titanium dioxide** is used as a pigment in white paint because it reflects visible light very strongly.

a) i) What size would you expect the titanium dioxide particles in the paint to be?

A 1-100 nm **B** 0.1-5 μm **C** 10-100 mm

ii) Give a reason for your choice.

....

....

b) Nanoparticles of titanium dioxide are used in sunscreens. Match the properties of titanium dioxide nanoparticles to their benefit in sunscreen.

Properties	Benefits
absorb visible light	leaves no marks on the skin
reflect UV light	not dissolved by sweat
insoluble in water	prevents harmful rays reaching the skin

Q4 In an episode of the TV series 'Star Trek — The Next Generation', two **nanites** (robots smaller than living cells) escape from a genetics experiment and enter the computer system. They begin to **replicate** rapidly, **destroying** all the starship's essential systems.

a) How do real nanomachines differ from the fictional nanites described in the TV series?

....

b) Explain the advantages of developing nanomachines.

....

....

Relative Formula Mass

Q1 a) What is meant by the **relative atomic mass** of an element?

....

b) What are the **relative atomic masses (A_r)** of the following:

i) magnesium .... **iv)** hydrogen .... **vii)** K ....

ii) neon .... **v)** C .... **viii)** Ca ....

iii) oxygen .... **vi)** Cu .... **ix)** Cl ....

Q2 **Identify** the **elements** A, B and C.

Element A has an A_r of 4.
Element B has an A_r 3 times that of element A.
Element C has an A_r 4 times that of element A.

Element A is ....

Element B is ....

Element C is ....

Q3 What are the **relative formula masses (M_r)** of the following:

a) water (H_2O) ....

b) potassium hydroxide (KOH) ....

c) nitric acid (HNO_3) ....

d) sulfuric acid (H_2SO_4) ....

e) ammonium nitrate (NH_4NO_3) ....

f) aluminium sulfate ($Al_2(SO_4)_3$) ....

Q4 The equation below shows a reaction between an element, X, and water. The total M_r of the products is **114**. What is substance X?

$$2X + 2H_2O \rightarrow 2XOH + H_2$$

....

....

Top Tips: The **periodic table** really comes in useful here. There's no way you'll be able to answer these questions without one (unless you've memorised all the elements' relative atomic masses — and that would just be **silly**). Luckily for you, you'll be **given** everything you need in your **exam**.

Isotopes and Relative Atomic Mass

Q1 Choose the correct words to complete this paragraph.

electrons	element	isotopes	protons	compound	neutrons

.... are different atomic forms of the same .... which have the same number of .... but different numbers of ....

Q2 Which of the following atoms are **isotopes** of each other? Explain how you know.

W ${}^{12}_{6}\mathrm{C}$ **X** ${}^{40}_{18}\mathrm{Ar}$ **Y** ${}^{14}_{6}\mathrm{C}$ **Z** ${}^{40}_{20}\mathrm{Ca}$

.... and .... because ....

Q3 **Carbon-14** is an unstable isotope of carbon.

a) How many of the following particles does one atom of carbon-14 contain?

i) neutrons .... **ii)** protons .... **iii)** electrons ....

b) Carbon-12 is the more common isotope of carbon.
Would you expect it to have different chemical properties from carbon-14? Explain your answer.

....

Q4 Draw lines to join the beginning of each sentence to its correct ending.

Relative atomic mass is

Relative abundance means

the proportion of one isotope in an element

the average mass of all atoms of that element

Q5 **Chlorine** has two main **isotopes**, ${}^{35}\mathrm{Cl}$ and ${}^{37}\mathrm{Cl}$. Their relative abundances are shown in the table.

relative mass of isotope	relative abundance
35	3
37	1

Use this information to calculate the relative atomic mass of chlorine.

....

....

Percentage Mass and Empirical Formulas

Q1 a) Write down the **formula** for calculating the **percentage mass** of an element in a compound.

....

b) Calculate the percentage mass of the following elements in ammonium nitrate, NH_4NO_3.

i) Nitrogen ....

ii) Hydrogen ....

iii) Oxygen ....

Q2 **Nitrogen monoxide**, NO, reacts with oxygen, O_2, to form **oxide R**.

a) Calculate the percentage mass of nitrogen in **nitrogen monoxide**.

....

b) Oxide R has a percentage composition by mass of **30.4% nitrogen** and **69.6% oxygen**. Work out its empirical formula.

....

....

....

Q3 1.48 g of a **calcium compound** contains 0.8 g of calcium, 0.64 g of oxygen and 0.04 g of hydrogen.

Work out the empirical formula of the compound.

....

....

....

Q4 a) Calculate the percentage mass of **oxygen** in each of the following compounds.

A Fe_2O_3 **B** H_2O **C** $CaCO_3$

....

....

....

b) Which compound has the **greatest** percentage mass of oxygen? ....

Calculating Masses in Reactions

Q1 Anna burns **10 g** of **magnesium** in air to produce **magnesium oxide** (MgO).

a) Write out the **balanced equation** for this reaction.

b) Calculate the mass of **magnesium oxide** that's produced.

Q2 What mass of **sodium** is needed to make **2 g** of **sodium oxide**? $4Na + O_2 \rightarrow 2Na_2O$

Q3 **Aluminium** and **iron oxide** (Fe_2O_3) react together to produce **aluminium oxide** (Al_2O_3) and **iron.**

a) Write out the **balanced equation** for this reaction.

b) What **mass** of iron is produced from **20 g** of iron oxide?

Q4 When heated, **limestone** ($CaCO_3$) decomposes to form **calcium oxide** (CaO) and **carbon dioxide**.

How many **kilograms** of limestone are needed to make **100 kilograms** of **calcium oxide**?

The calculation is the same — just use 'kg' instead of 'g'.

Calculating Masses in Reactions

Q5 **Iron oxide** is reduced to **iron** inside a blast furnace using carbon. There are **three** stages involved.

Stage A	$C + O_2 \rightarrow CO_2$
Stage B	$CO_2 + C \rightarrow 2CO$
Stage C	$3CO + Fe_2O_3 \rightarrow 2Fe + 3CO_2$

a) If **10 g** of **carbon** are used in stage B, and all the carbon monoxide produced gets used in stage C, what **mass** of CO_2 is produced in **stage C**?

Work out the mass of CO at the end of stage B first.

....

....

....

....

b) Suggest what happens to the CO_2 produced in stage C.

Look at where CO_2 is used.

....

Q6 **Sodium sulfate** (Na_2SO_4) is made by reacting **sodium hydroxide** (NaOH) with **sulfuric acid** (H_2SO_4). **Water** is also produced.

a) Write out the **balanced equation** for this reaction.

....

b) What mass of **sodium hydroxide** is needed to make **75 g** of **sodium sulfate**?

....

....

....

....

c) What mass of **water** is formed when **50 g** of **sulfuric acid** reacts with sodium sulfate?

....

....

....

....

Top Tips: Masses, equations, formulas — they can all seem a bit scary. But don't worry, practice makes perfect. And once you get the hang of them you'll wonder what all the fuss was about.

The Mole

Q1 a) **Complete** the following sentence.

> One mole of atoms or molecules of any substance will have a ....
>
> in grams equal to the .... for that substance.

b) What is the **mass** of each of the following?

i) 1 mole of copper ....

ii) 3 moles of chlorine **gas** ....

iii) 2 moles of nitric acid (HNO_3) ....

iv) 0.5 moles of calcium carbonate ($CaCO_3$) ....

Q2 a) Write down the formula for calculating the **number of moles in a given mass**.

....

b) How many **moles** are there in each of the following?

i) 20 g of calcium ....

ii) 112 g of sulfur ....

iii) 159 g of copper oxide (CuO) ....

c) Calculate the **mass** of each of the following.

i) 2 moles of sodium ....

ii) 0.75 moles of magnesium oxide (MgO) ....

iii) 0.025 moles of lead chloride ($PbCl_2$) ....

Q3 Ali adds **13 g** of zinc to **50 cm³** of hydrochloric acid. All of the zinc reacts.

$$Zn + 2HCl \rightarrow ZnCl_2 + H_2$$

a) How many moles of **zinc** were added?

....

b) How many moles of **hydrochloric acid** reacted?

....

Look at the symbol equation.

The Mole

Q4 a) Write down the formula for calculating the **number of moles in a solution**.

....

b) Use the formula to calculate the number of moles in:

There are 1000 cm^3 in 1 litre.

i) 50 cm^3 of a 2 M solution. ....

ii) 250 cm^3 of a 0.5 M solution. ....

iii) 550 cm^3 of a 1.75 M solution. ....

Q5 Work out the concentration of the following solutions in **mol/dm³**.

a) 0.05 moles of copper(II) sulfate dissolved in 500 cm^3 of water.

....

b) 0.2 moles of magnesium hydroxide dissolved in 250 cm^3 of water.

....

c) 0.03 moles of sodium hydroxide dissolved in 20 cm^3 of water.

....

Q6 Dr Burette adds **0.6 g** of sodium to water. Sodium hydroxide and hydrogen form. (All the sodium reacts.)

a) Write a **balanced symbol equation** for this reaction.

....

b) What mass of **hydrogen** is produced?

....

....

c) Calculate the mass of **sodium hydroxide** produced.

....

....

Top Tips: So, you already know that the mole is not just a small burrowing animal. Now you need to make sure that you can convert between moles and grams. But that's not all — make sure you learn the formula for calculating the number of moles and you'll soon be sailing through those exam questions.

Atom Economy

Q1 **Copper oxide** can be reduced to copper by heating it with carbon.

copper oxide + carbon → copper + carbon dioxide

$$2CuO + C \rightarrow 2Cu + CO_2$$

a) What is the useful product in this reaction? ....

b) Calculate the atom economy.

$$\text{atom economy} = \frac{\text{total } M_r \text{ of useful products}}{\text{total } M_r \text{ of reactants}} \times 100$$

....

....

c) What percentage of the starting materials are wasted?

....

Q2 It is important in industry to find the **best atom economy**.

a) Explain why. ....

....

....

b) What types of reaction have the highest atom economies? Give an example.

....

Q3 **Titanium** can be reduced from titanium chloride ($TiCl_4$) using magnesium or sodium.

a) Work out the atom economy for the reaction:

i) with magnesium: $TiCl_4 + 2Mg \rightarrow Ti + 2MgCl_2$ ....

....

ii) with sodium: $TiCl_4 + 4Na \rightarrow Ti + 4NaCl$ ....

....

b) Which one has the better atom economy? ....

Q4 **Chromium** can be extracted from its oxide (Cr_2O_3) using **aluminium**. The products of the reaction are **aluminium oxide** and **chromium**.

Calculate the atom economy of this reaction.

....

....

Percentage Yield

Q1 James wanted to produce **silver chloride** (AgCl). He added a carefully measured mass of silver nitrate to an excess of dilute hydrochloric acid.

a) Write down the formula for calculating the **percentage yield** of a reaction.

....

b) James calculated that he should get 2.7 g of silver chloride, but he only got 1.2 g. What was the **percentage yield**?

....

Q2 Aaliya and Natasha mixed together barium chloride ($BaCl_2$) and sodium sulfate (Na_2SO_4) in a beaker. An **insoluble** substance formed. They **filtered** the solution to obtain the solid substance, and then transferred the solid to a clean piece of **filter paper** and left it to dry.

a) Aaliya calculated that they should produce a yield of **15 g** of barium sulfate. However, after completing the experiment they found they had only obtained **6 g**.

Calculate the **percentage yield** for this reaction.

....

b) Suggest two reasons why their actual yield was lower than their predicted yield.

1. ....

....

2. ....

....

Q3 The reaction between magnesium and oxygen produces a white powder, **magnesium oxide**. Four samples of magnesium, each weighing 2 g, were burned and the oxide produced was weighed. The **expected** yield was **3.33 g**.

Sample	Mass of oxide (g)
A	3.00
B	3.18
C	3.05
D	3.15

a) What is the percentage yield for each sample?

....

....

....

b) Which of the following are likely reasons why the yield was not 100%? Circle their letters.

A The reaction was too fast

B Too much magnesium was burned

C The magnesium was not pure

D Some of the oxide was lost before it was weighed

Acids and Bases

Q1 a) Complete the equation below for the reaction between an **acid** and a **base**.

acid + base → .... + ...

b) Circle the correct term for this kind of reaction.

decomposition **oxidation** **neutralisation**

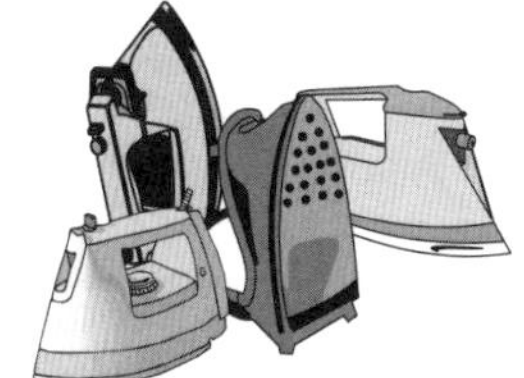

c) Which of the following ions:

$H^+(aq)$	$OH^-(aq)$	$Cl^-(aq)$	$Na^+(aq)$

i) react with each other to form water? ....

ii) is present in an acidic solution? ....

iii) is present in an alkaline solution? ....

iv) would combine to form the salt sodium chloride? ....

v) would be present in a solution with a pH of 10? ....

vi) would be found in lemon juice? ....

Q2 Joey wanted to test whether some antacid tablets really do **neutralise acid**.

He added a tablet to some hydrochloric acid, stirred it until it dissolved and tested the pH of the solution. Further tests were carried out after dissolving a second, third and fourth tablet. His results are shown in the table below.

Number of Tablets	pH
0	1
1	2
2	3
3	7
4	9

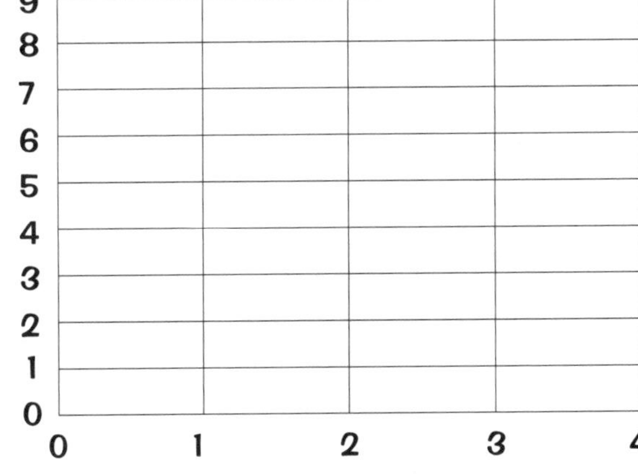

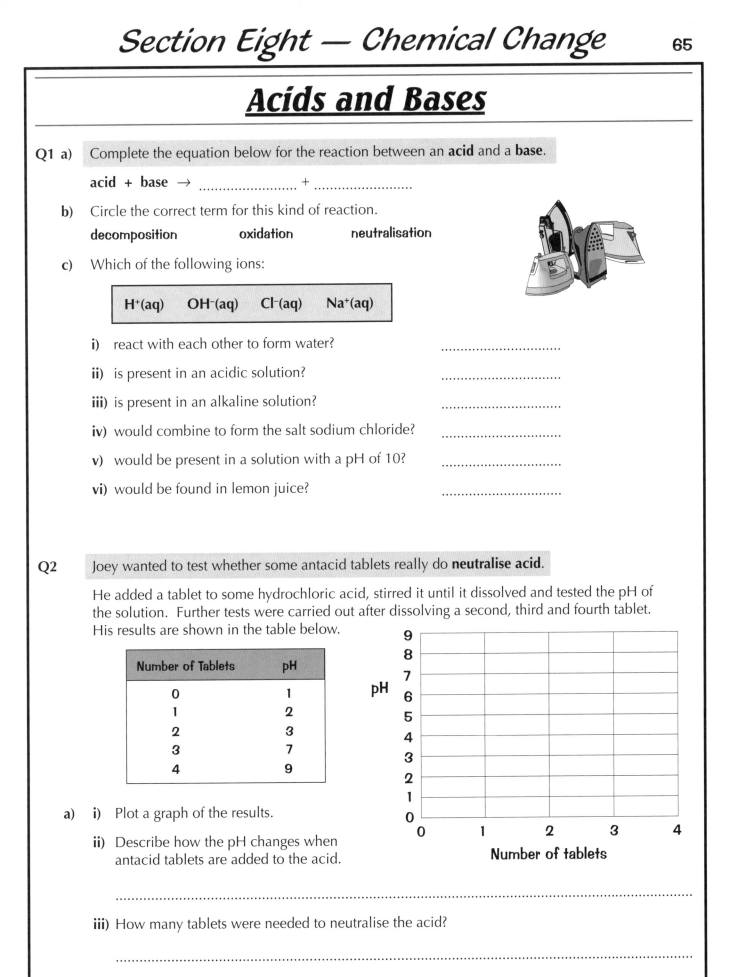

a) i) Plot a graph of the results.

ii) Describe how the pH changes when antacid tablets are added to the acid.

....

iii) How many tablets were needed to neutralise the acid?

....

b) Joey tested another brand of tablets and found that **two** tablets neutralised the same volume of acid. On the graph, sketch the results you might expect for these tablets.

Acids and Bases

Q3 Complete the following sentences.

a) Solutions which are not acidic or alkaline are said to be ....

b) Universal indicator is a combination of different coloured ....

c) If a substance is neutral it has a pH of .... A solution of this substance would turn .... with universal indicator.

Q4 Ants' stings hurt because of the **formic acid** they release. The pH measurements of some household substances are given in the table.

SUBSTANCE	pH
lemon juice	4
baking soda	9
caustic soda	14
soap powder	11

a) Describe how you could test the formic acid to find its pH value.

....

....

b) Suggest a substance from the list that could be used to relieve the discomfort of an ant sting. Explain your answer.

....

....

c) Explain why universal indicator only gives an **estimate** of the pH.

....

....

Q5 Modern industry uses thousands of tonnes of **sulfuric acid** per day.

a) Give two uses of sulfuric acid in the car manufacturing industry.

1. .... 2. ....

b) Which of the following compounds found in fertilisers is manufactured from sulfuric acid?

ammonium nitrate **ammonium sulfate** **ammonium phosphate** **potassium nitrate**

c) Describe how sulfuric acid is used in the preparation of metal surfaces.

....

....

Top Tips: Ahh... acids and bases. They pop up everywhere. And I mean EVERYWHERE. The chemistry lab, the human body, vehicles, poisons and antidotes, even in the kitchen sink.

Acids Reacting with Metals

Q1 Fill in the blanks using some of the words given below.

reactive silver nitric more hydrogen less chlorides
sulfuric carbon dioxide non-metals nitrates metals

Acids react with most .... to form salts and .... gas.
Metals like copper and .... which are less .... than hydrogen don't react with acids. The .... reactive the metal, the more vigorously the bubbles of gas form. Hydrochloric acid forms .... and .... acid produces sulfates. However, the reactions of metals with .... acid don't follow this simple pattern.

Q2 Rhiannon is planning an experiment to investigate the rate of reaction between magnesium and different **concentrations** of hydrochloric acid.

a) How could she measure the rate of the reaction?

....

b) What is the **independent variable** in her experiment?

....

c) Give two variables that she will need to keep the **same** in her experiment.

....

Q3 a) Write out the **balanced** versions of the following equations.

i) $Ca + HCl \rightarrow CaCl_2 + H_2$ ....

ii) $Na + HCl \rightarrow NaCl + H_2$ ....

iii) $Li + H_2SO_4 \rightarrow Li_2SO_4 + H_2$ ....

b) Hydrobromic acid reacts with magnesium as shown in the equation below to form a bromide salt and hydrogen.

$$Mg + 2HBr \rightarrow MgBr_2 + H_2$$

i) Name the salt formed in this reaction. ....

ii) Write a balanced symbol equation for the reaction between aluminium and hydrobromic acid. (The formula of aluminium bromide is $AlBr_3$.)

....

Acids Reacting with Metals

Q4 The diagram below shows **aluminium** reacting with **sulfuric acid**.

a) Label the diagram with the names of the chemicals.

....

....

....

b) Complete the word equation for this reaction:

aluminium + ... → **aluminium sulfate +** ....

c) Write a balanced symbol equation for this reaction.

....

The formula of aluminium sulfate is $Al_2(SO_4)_3$.

d) Zinc also reacts with sulfuric acid. Give the word equation for this reaction.

....

e) Write a balanced symbol equation for the reaction between:

i) magnesium and hydrochloric acid ....

ii) calcium and nitric acid ....

Q5 The table shows what happens when different **metals** react with **hydrochloric acid**.

Metal	A	B	C	D
Observations	gas bubbles formed vigorously metal dissolved quickly	no gas bubbles formed metal unaffected by the acid	gas bubbles form slowly most of the metal remained after 5 min	gas bubbles form steadily most of the metal dissolved after 5 min

a) Which is the **most** reactive metal? ....

b) Which metal(s) are **less** reactive than hydrogen? ....

c) The metals used in this experiment were magnesium, zinc, iron and copper. Match each of these metals to the correct letter from the table.

A ....

B ....

C ....

D ....

Look at the reactivity series to help you.

most reactive	magnesium
	carbon
	zinc
	iron
	lead
	copper
least reactive	gold

Neutralisation Reactions

Q1 Fill in the blanks to complete the word equations for **acids** reacting with **metal oxides** and **metal hydroxides**.

a) hydrochloric acid + lead oxide → .... chloride + water

b) nitric acid + copper hydroxide → copper .... + water

c) sulfuric acid + zinc oxide → zinc sulfate + ...

d) hydrochloric acid + ... oxide → nickel .... + ...

e) .... acid + copper oxide → .... nitrate + ...

f) sulfuric acid + ... hydroxide → sodium .... + ...

Q2 a) Put a tick in the box next to any of the sentences below which are **true**.

Alkalis are bases which don't dissolve in water. ☐

Acids react with metal oxides to form a salt and water. ☐

Hydrogen gas is formed when an acid reacts with an alkali. ☐

Salts and water are formed when acids react with metal hydroxides. ☐

Ammonia solution is alkaline. ☐

Calcium hydroxide is an acid that dissolves in water. ☐

b) Use the formulas below to write **symbol equations** for two acid/base reactions.

H_2SO_4 H_2O H_2O CuO $NaCl$ HCl $CuSO_4$ $NaOH$

....

....

Q3 Name two substances which would react to make each of the following **salts**.

a) Potassium sulfate ....

b) Ammonium chloride ....

c) Silver nitrate ....

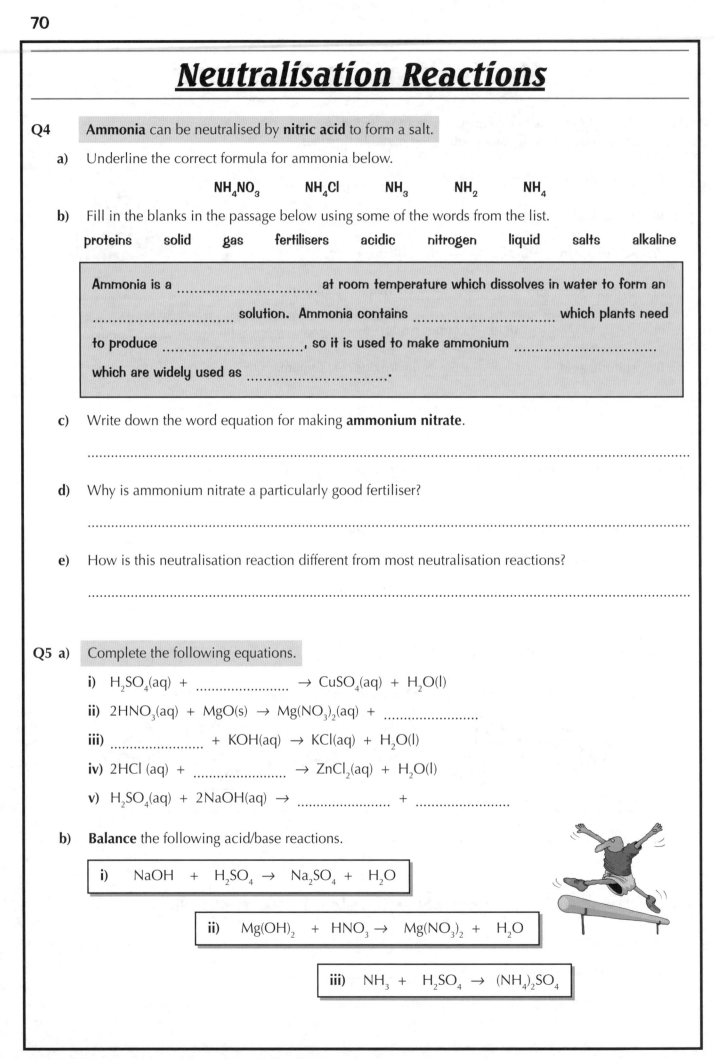

Neutralisation Reactions

Q4 **Ammonia** can be neutralised by **nitric acid** to form a salt.

a) Underline the correct formula for ammonia below.

NH_4NO_3 NH_4Cl NH_3 NH_2 NH_4

b) Fill in the blanks in the passage below using some of the words from the list.

proteins solid gas fertilisers acidic nitrogen liquid salts alkaline

> Ammonia is a .... at room temperature which dissolves in water to form an .... solution. Ammonia contains .... which plants need to produce ...., so it is used to make ammonium .... which are widely used as ....

c) Write down the word equation for making **ammonium nitrate**.

....

d) Why is ammonium nitrate a particularly good fertiliser?

....

e) How is this neutralisation reaction different from most neutralisation reactions?

....

Q5 a) Complete the following equations.

i) $H_2SO_4(aq)$ + ... $\rightarrow CuSO_4(aq) + H_2O(l)$

ii) $2HNO_3(aq) + MgO(s) \rightarrow Mg(NO_3)_2(aq)$ + ...

iii) .... + $KOH(aq) \rightarrow KCl(aq) + H_2O(l)$

iv) $2HCl\ (aq)$ + ... $\rightarrow ZnCl_2(aq) + H_2O(l)$

v) $H_2SO_4(aq) + 2NaOH(aq) \rightarrow$ .... + ...

b) **Balance** the following acid/base reactions.

i) $NaOH + H_2SO_4 \rightarrow Na_2SO_4 + H_2O$

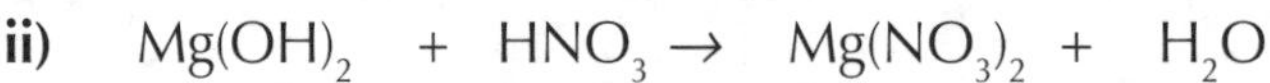
ii) $Mg(OH)_2 + HNO_3 \rightarrow Mg(NO_3)_2 + H_2O$

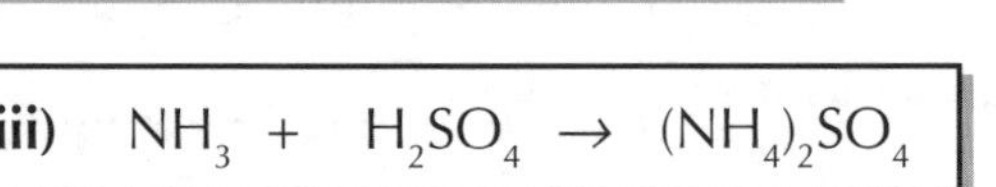
iii) $NH_3 + H_2SO_4 \rightarrow (NH_4)_2SO_4$

Making Salts

Q1 Complete the following sentences by circling the correct word from each pair.

a) Most chlorides, sulfates and nitrates are **soluble** / **insoluble** in water.

b) Most oxides, hydroxides and carbonates are **soluble** / **insoluble** in water.

c) Soluble salts can be made by reacting acids with **soluble** / **insoluble** bases until they are just **neutralised** / **displaced**.

Care of Animals
Rule No. 57:
Never pour salt in
a rabbit's eyes.

d) Insoluble salts are made by **precipitation** / **electrolysis**.

e) Salts can be made by displacement, where a **more** / **less** reactive metal is put into a salt solution of a **more** / **less** reactive metal.

Q2 **A**, **B**, **C** and **D** are symbol equations for reactions in which **salts** are formed.

A $CuO(s) + H_2SO_4(aq) \rightarrow CuSO_4(aq) + H_2O(l)$

B $2NaOH(aq) + H_2SO_4(aq) \rightarrow Na_2SO_4(aq) + 2H_2O(l)$

C $Zn(s) + 2AgNO_3(aq) \rightarrow Zn(NO_3)_2(aq) + 2Ag(s)$

D $Pb(NO_3)_2(aq) + H_2SO_4(aq) \rightarrow PbSO_4(s) + 2HNO_3(aq)$

Which equation (A, B, C or D) refers to the formation of a salt:

a) in an acid/alkali reaction. ....

b) by precipitation. ....

c) from an insoluble base. ....

d) by displacement. ....

Q3 A piece of **magnesium** is dropped into a solution of **copper sulfate**.

a) Explain why the piece of magnesium becomes coated in an **orange-coloured** substance.

....

....

b) A new salt is formed during this experiment. Name this salt.

....

c) Once all the magnesium is coated with the orange substance, the reaction stops. Explain why.

....

....

d) What type of reaction is this? Circle your answer. **displacement** **neutralisation** **electrolysis**

Making Salts

Q4 **Silver chloride** is an insoluble salt which is formed as a **precipitate** when silver nitrate and sodium chloride solutions are mixed together.

a) Complete the word equation for the reaction.

.... + ... → silver chloride + ...

b) After mixing the solutions to produce a precipitate, what further steps are needed to produce a dry sample of silver chloride?

....

....

Q5 **Nickel sulfate** (a soluble salt) was made by adding insoluble **nickel carbonate** to **sulfuric acid** until no further reaction occurred.

nickel carbonate

acid

excess nickel carbonate

a) What piece of apparatus is being used to add the nickel carbonate in the diagram?

....

b) State two observations that would tell you that the reaction was complete.

....

....

Once the reaction was complete, the excess nickel carbonate was separated from the nickel sulfate solution using the apparatus shown.

c) Label the diagram which shows the separation process.

....

....

....

d) What is this method of separation called?

....

e) Describe how you could produce a solid sample of nickel sulfate from nickel sulfate solution.

....

....

More Chemical Changes

Q1 Answer the following questions about reactions involving **water**.

a) Which reaction(s) involve **dehydration**? Tick one or more boxes.

- $Cu(OH)_2\ (s) \rightarrow CuO\ (s) + H_2O\ (l)$ ☐
- $C_2H_4\ (g) + H_2O\ (l) \rightarrow C_2H_5OH\ (l)$ ☐
- $NiCO_3\ (s) \rightarrow NiO\ (s) + CO_2\ (g)$ ☐
- ethanol + ethanoic acid → ethyl ethanoate + water ☐

b) Which reaction(s) involve **hydration**? Tick one or more boxes.

- $CuSO_4.5H_2O\ (s) \rightarrow CuSO_4\ (s) + 5H_2O\ (l)$ ☐
- $CaO\ (s) + H_2O\ (l) \rightarrow Ca(OH)_2\ (s)$ ☐
- $CH_4\ (g) + 3O_2\ (g) \rightarrow CO_2\ (g) + 2H_2O\ (l)$ ☐
- $C_6H_{12}O_6\ (s) \rightarrow 6C\ (s) + 6H_2O\ (l)$ ☐

Q2 Tick the correct boxes to show whether the following statements are **true** or **false**.

		True	False
a)	**i)** Calcium carbonate gives off carbon dioxide and water when it decomposes.	☐	☐
	ii) Thermal decomposition involves breaking a substance down into simpler substances.	☐	☐
	iii) Condensation reactions are a type of dehydration reaction.	☐	☐
	iv) Sugar can be hydrated by adding it to concentrated sulfuric acid.	☐	☐
		☐	☐

b) Write a word equation for the thermal decomposition of baking powder.

....

Q3 When **hydrated copper sulfate** (blue crystals) is heated it turns into **anhydrous copper sulfate** (white powder) and **water**.

$$CuSO_4.5H_2O(s) \rightarrow CuSO_4(s) + 5H_2O(l)$$

a) Why is the reaction described as dehydration?

....

b) Why can the reaction also be described as thermal decomposition?

....

c) Why is the reverse reaction described as hydration?

....

More Chemical Changes

Q4 Clear, blue **copper(II) sulfate solution** and clear, colourless **sodium hydroxide** solution were mixed. The liquid went cloudy and pale blue. After a while a **pale blue solid** was left at the bottom and the liquid was **clear** again.

a) What type of reaction has occurred? ....

b) Name the blue solid formed. ....

c) Write a balanced symbol equation for this reaction.

....

Q5 Cilla adds a few drops of **NaOH** solution to solutions of different **metal compounds**.

a) Complete her table of results.

Compound	Metal Cation	Colour of Precipitate
copper(II) sulfate		blue
iron(II) sulfate		
iron(III) chloride	Fe^{3+}	
copper(II) chloride		

b) Write a balanced symbol equation for the reaction of copper(II) chloride with sodium hydroxide.

....

Q6 The diagram shows an experiment into **rusting**.

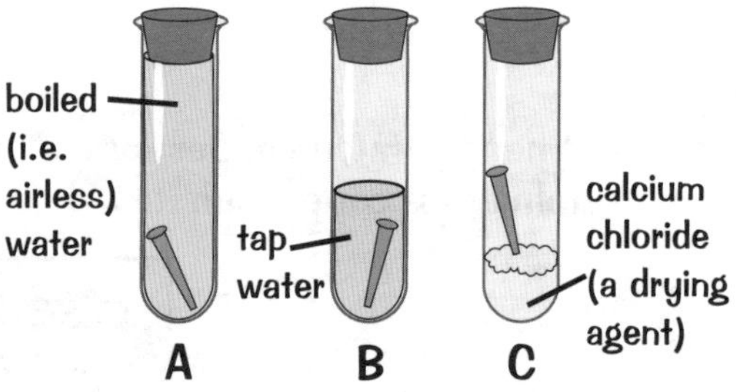

a) In which tube will the iron nail rust most quickly?

....

b) This experiment can be used to demonstrate that two substances are needed for iron to rust. What are they?

....

boiled (i.e. airless) water

tap water

calcium chloride (a drying agent)

A B C

c) Write a word equation for rusting. (Make sure you use the correct scientific term for rust.)

....

d) Explain why people who live near to the sea should wash their cars frequently.

....

....

Electrolysis and the Half-Equations

Q1 Fill in the blanks to complete the paragraph below.

Electrolysis is breaking down a substance using ....
It needs to happen in a conducting liquid called the ....
During electrolysis, ... are taken away from ions at the positive electrode, called the ...., and given to other ions at the negative electrode, called the ....

Q2 The diagram below shows the electrolysis of a **salt solution**.

a) Identify the ions and molecules labelled A, B, C and D on the diagram.

Choose from the options in the box.

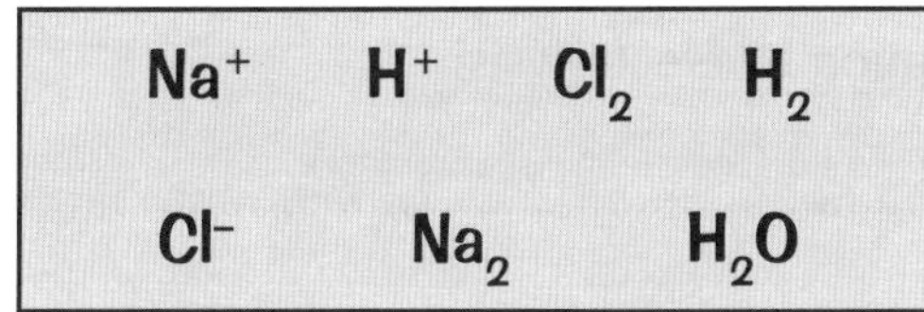

Na^+	H^+	Cl_2	H_2
Cl^-	Na_2	H_2O	

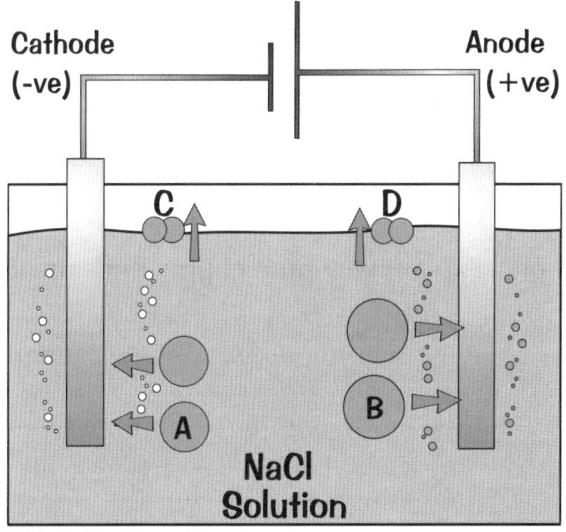

A .... B ....

C .... D ....

b) Write **balanced** half-equations for the processes that occur during the electrolysis of this salt solution.

Cathode: ....

Anode: ....

Q3 Explain why a substance needs to be either in a **solution** or **molten** for electrolysis to work.

....

....

Top Tips: Half-equations just show what's going on at the cathode and anode in terms of electrons — a positive ion gains electrons (and a negative ion loses electrons) to make neutral atoms.

Electrolysis and the Half-Equations

Q4 Write **half-equations** for the purification of copper by electrolysis. Include state symbols.

Cathode: ....

Anode: ....

Q5 During the electrolytic purification of copper, the **impure sludge** simply falls to the bottom. It does **not** follow the copper ions to the cathode. Why do you think this is?

....

....

The copper ions that leave the anode are positively charged.

Q6 **Copper** is extracted from its ore by **reduction** with **carbon**.

a) Why might copper extracted from the ground need to be purified?

....

b) What is used as the **anode** during copper purification by electrolysis?

....

c) Explain why pure copper ends up at the **cathode** during electrolysis.

....

....

....

Q7 Silver can be purified in the same way as copper. Write **half-equations** for the processes that takes place at the anode and the cathode.

Silver forms 1^+ ions.

Cathode: ....

Anode: ....

Q8 Why would it **not** be a good idea to carry out the electrolysis of **copper** in an electrolyte that contained **zinc** ions instead of copper ions? Tick the correct box.

The zinc ions will not conduct an electrical current. ☐

The copper produced will have zinc impurities in it. ☐

A poisonous gas would be produced. ☐

The zinc and copper ions will react with each other. ☐

Rates of Reaction

Q1 Circle the correct words to complete the statements below about **rates of reaction**.

a) The higher the temperature, the **faster** / **slower** the rate of a reaction.

b) A higher concentration will **increase** / **reduce** the rate of a reaction.

c) If the reactants are **gases** / **liquids**, a higher pressure will give a **faster** / **slower** reaction.

d) A larger particle size **increases** / **decreases** the rate of reaction.

e) A catalyst **speeds up** / **slows down** the rate of reaction and **is** / **isn't** used up.

Q2 In an experiment, **different sizes** of marble chips were reacted with excess hydrochloric acid. The **same mass** of marble was used each time. The graph below shows how much **gas** was produced when using large marble chips (X), medium marble chips (Y) and small marble chips (Z).

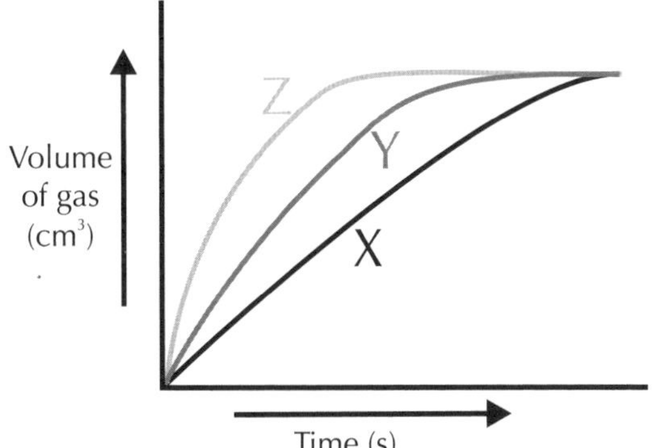

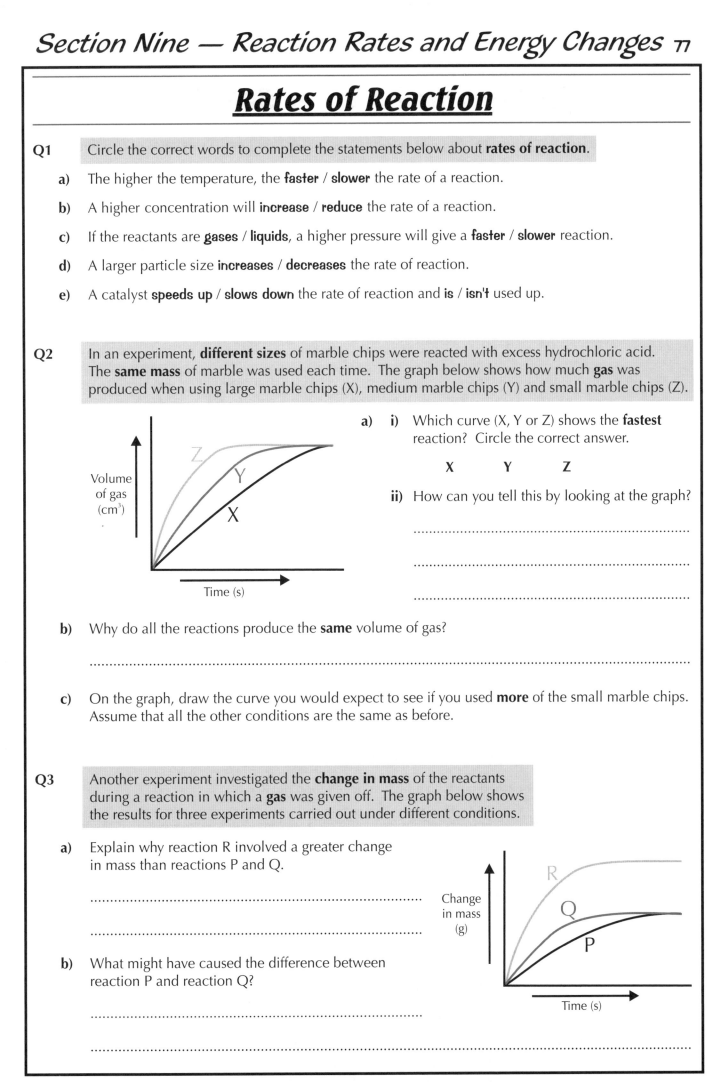

a) i) Which curve (X, Y or Z) shows the **fastest** reaction? Circle the correct answer.

X **Y** **Z**

ii) How can you tell this by looking at the graph?

....

....

....

b) Why do all the reactions produce the **same** volume of gas?

....

c) On the graph, draw the curve you would expect to see if you used **more** of the small marble chips. Assume that all the other conditions are the same as before.

Q3 Another experiment investigated the **change in mass** of the reactants during a reaction in which a **gas** was given off. The graph below shows the results for three experiments carried out under different conditions.

a) Explain why reaction R involved a greater change in mass than reactions P and Q.

....

....

Change in mass (g)
Time (s)
R
Q
P

b) What might have caused the difference between reaction P and reaction Q?

....

....

Measuring Rates of Reaction

Q1 Use the words provided to complete the sentences below about **measuring rates of reaction**.

faster	speed	volume	reactants	gas	mass	formed	precipitation

a) The .... of a reaction can be measured by observing either how quickly the .... are used up or how quickly the products are ....

b) In a .... reaction you usually measure how quickly the product is formed. The product turns the solution cloudy. The .... it turns cloudy the faster the reaction.

c) In a reaction that produces a .... you can measure how quickly the .... of the reactants changes or measure the .... given off in a certain time interval.

Q2 Sam conducted two experiments with equal masses of marble chips and equal volumes of hydrochloric acid (HCl). He used two **different concentrations** of acid and measured the **change in mass** of the reactants. Below is a graph of the results.

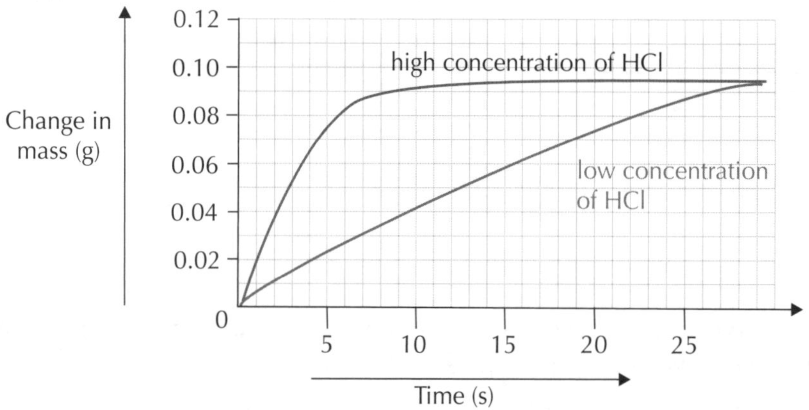

a) Calculate the average rate of reaction for the first five seconds for:

i) the experiment carried out with a high concentration of HCl.

....

ii) the experiment carried out with a low concentration of HCl.

....

b) Circle the letter(s) to show the valid conclusion(s) you might draw **from this graph**.

A Rate of reaction depends on the temperature of the reactants.

B Increasing the concentration of the acid has no effect on the rate of reaction.

C Rate of reaction depends on the acid concentration.

D Rate of reaction depends on the mass of the marble chips.

Measuring Rates of Reaction

Q3 Charlie was comparing the rates of reaction of 5 g of magnesium ribbon with 20 cm^3 of **five different concentrations** of hydrochloric acid. Each time he measured how much **gas** was produced during the **first minute** of the reaction. He did the experiment **twice** for each concentration of acid and obtained these results:

Concentration of HCl (mol/dm^3)	Experiment 1 — volume of gas produced (cm^3)	Experiment 2 — volume of gas produced (cm^3)	Average volume of gas produced (cm^3)
2	92	96	
1.5	63	65	
1	44	47	
0.5	20	50	
0.25	9	9	

a) Fill in the last column of the table.

b) Circle the **anomalous** result in the table.

c) Which concentration of hydrochloric acid produced the fastest rate of reaction? ....

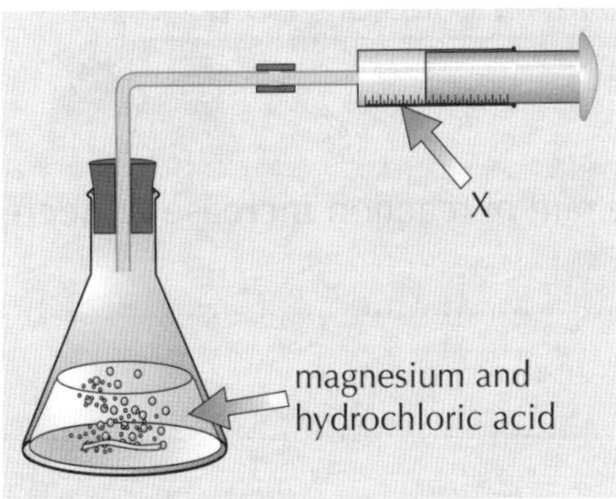
X
magnesium and hydrochloric acid

d) A diagram of the **apparatus** used in the experiment is shown on the left.

i) What is the object marked **X** called?

....

ii) Name one other key piece of apparatus needed for this experiment (not shown in the diagram).

....

e) **Sketch** a graph of the average volume of gas produced against concentration of HCl and label the axes. Do not include the anomalous result.

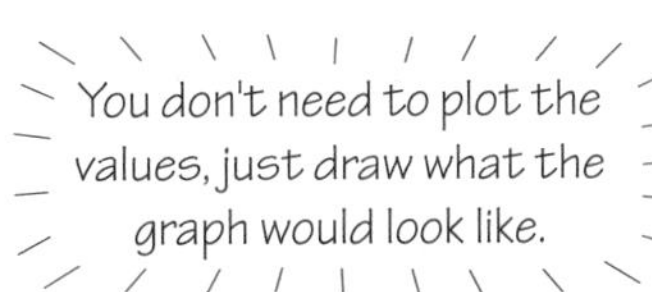
You don't need to plot the values, just draw what the graph would look like.

f) Why did Charlie do the experiment twice and calculate the average volume?

....

g) How might the anomalous result have come about?

....

h) Suggest two changes Charlie could make to improve his results if he repeated his investigation.

1. ....

2. ....

Collision Theory

Q1 Circle the correct words to complete the sentences.

a) In order for a reaction to occur, the particles must **remain still / collide**.

b) If you heat up a reaction mixture, you give the particles more **energy / surface area**.

c) This makes them move **faster / more slowly**, so there is **more / less** chance of successful collisions.

d) So, increasing the temperature increases the **concentration / rate** of reaction.

Q2 Draw lines to match up the **changes** with their **effects**.

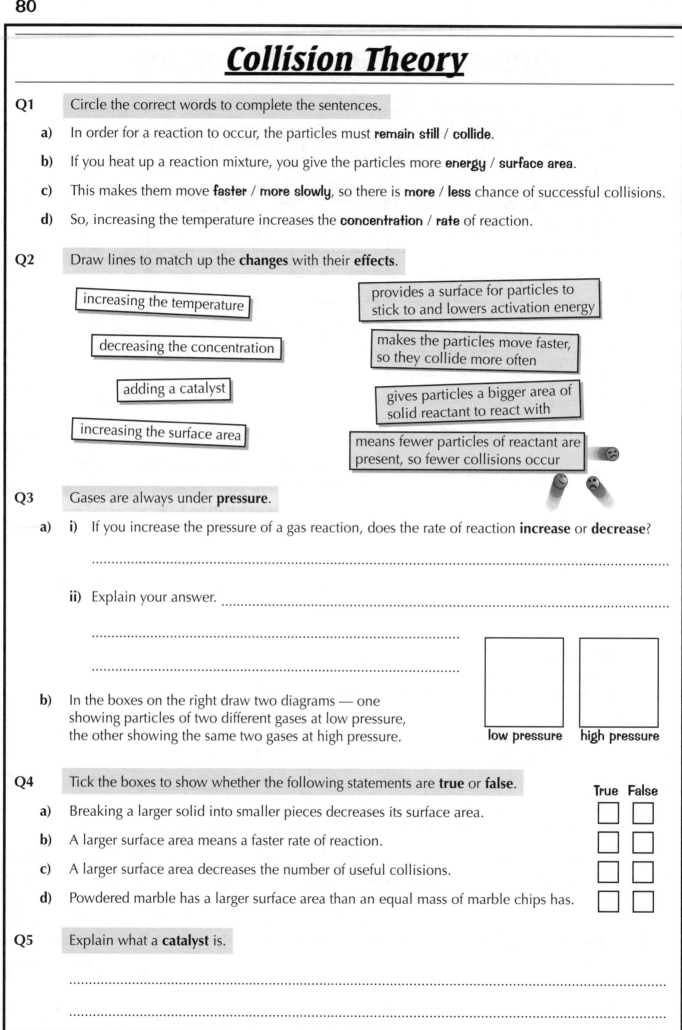

Changes	Effects
increasing the temperature	provides a surface for particles to stick to and lowers activation energy
decreasing the concentration	makes the particles move faster, so they collide more often
adding a catalyst	gives particles a bigger area of solid reactant to react with
increasing the surface area	means fewer particles of reactant are present, so fewer collisions occur

Q3 Gases are always under **pressure**.

a) i) If you increase the pressure of a gas reaction, does the rate of reaction **increase** or **decrease**?

....

ii) Explain your answer. ....

....

....

b) In the boxes on the right draw two diagrams — one showing particles of two different gases at low pressure, the other showing the same two gases at high pressure.

low pressure　　high pressure

Q4 Tick the boxes to show whether the following statements are **true** or **false**.

		True	False
a)	Breaking a larger solid into smaller pieces decreases its surface area.	☐	☐
b)	A larger surface area means a faster rate of reaction.	☐	☐
c)	A larger surface area decreases the number of useful collisions.	☐	☐
d)	Powdered marble has a larger surface area than an equal mass of marble chips has.	☐	☐

Q5 Explain what a **catalyst** is.

....

....

Catalysts

Q1 To get a reaction to **start**, you have to give the particles some **energy**.

a) What is this energy called? Underline the correct answer.

potential energy **activation energy** **chemical energy**

b) The diagram opposite shows two reactions — one with a catalyst and one without. Which line shows the reaction **with** a catalyst?

....

c) On this diagram, draw and label arrows to show the activation energy for the reaction without a catalyst and the activation energy for the reaction with a catalyst.

Energy
B
A
Reactants
Products
Progress of Reaction

Q2 Solid catalysts come in **different forms**. Two examples are **pellets** and **fine gauze**.

Explain why solid catalysts are used in forms such as these.

....

....

Q3 Industrial catalysts are often **metals**.

You find them in the middle of the periodic table.

a) Which type of metals are commonly used?

....

b) Give an example of a metal catalyst and say which industrial process it is used in.

....

Q4 Catalysts are very important in **industrial reactions**.

a) Give two ways in which using catalysts can save money for a manufacturer.

1. ....

2. ....

b) Give two possible disadvantages of using a catalyst in an industrial reaction.

1. ....

2. ....

Energy Transfer in Reactions

Q1 Use the words below to complete the blanks in the passage. Each word can be used more than once.

endothermic **exothermic** **energy** **heat** **an increase** **a decrease**

All chemical reactions involve changes in ....

In .... reactions, ... energy is given out to the surroundings. A thermometer will show .... in temperature.

In .... reactions, ... energy is taken in from the surroundings. A thermometer will show .... in temperature.

Q2 Fiz investigated the **temperature change** during a reaction. She added 25 cm^3 of sodium hydroxide solution to 25 cm^3 of hydrochloric acid. She used a **data logger** to measure the temperature of the reaction over the first **five** seconds.

Fiz plotted her results on the graph shown.

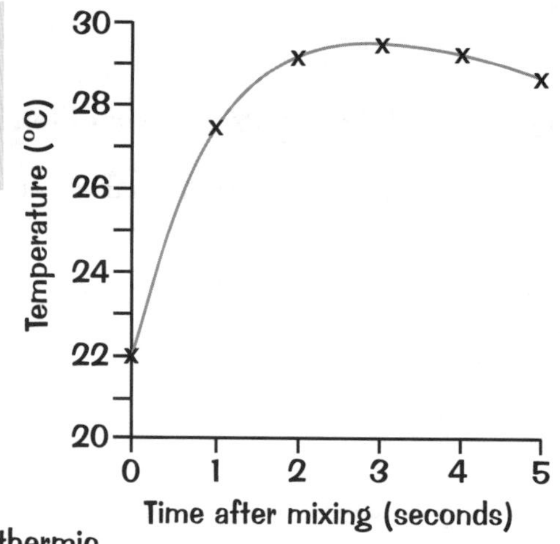

a) What was the increase in temperature due to the reaction?

....

b) Circle any of the words below that correctly describe the reaction in this experiment.

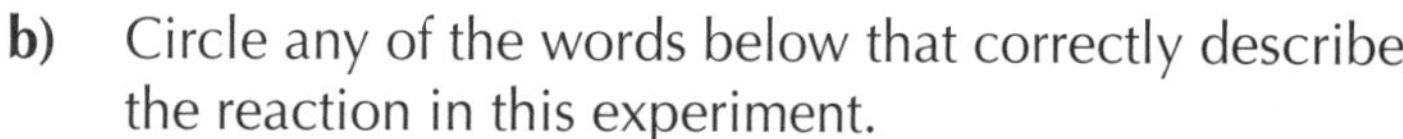

Q3 State whether bond **breaking** and bond **forming** are exothermic or endothermic reactions, and explain why in both cases.

Bond breaking ....

....

Bond forming ....

....

Q4 Decomposing 1 tonne (1000 kg) of $CaCO_3$ requires about 1800 000 kJ of **heat energy**.

a) How much heat energy would be needed to make **1 kg** of $CaCO_3$ decompose?

....

b) How much $CaCO_3$ could be decomposed by **90 000 kJ** of heat energy?

....

Energy Transfer in Reactions

Q5 When **methane** burns in oxygen it forms carbon dioxide and water. The bonds in the methane and oxygen molecules **break** and new bonds are formed to make carbon dioxide and water molecules.

a) Is energy **taken in** or **given out** when the bonds in the methane and oxygen molecules **break**?

....

b) Is energy **taken in** or **given out** when the bonds in the carbon dioxide and water molecules **form**?

....

c) Methane is a fuel commonly used in cooking and heating. Do you think that burning methane is an exothermic or an endothermic process? Explain your answer.

....

....

d) Which of the following statements about burning methane is true? Circle one letter.

A **The energy involved in breaking bonds is greater than the energy involved in forming bonds.**

B **The energy involved in breaking bonds is less than the energy involved in forming bonds.**

C **The energy involved in breaking bonds is the same as the energy involved in forming bonds.**

Q6 Here are some practical uses of chemical reactions. Decide whether each reaction is **endothermic** or **exothermic**. In the box, put **N** for endothermic and **X** for exothermic.

a) A camping stove burns methylated spirit to heat a pan of beans. ☐

b) Special chemical cool packs are used by athletes to treat injuries. They are placed on the skin and draw heat away from the injury. ☐

c) Self-heating cans of coffee contain chemicals in the base. When the chemicals are combined they produce heat which warms the can. ☐

d) Baking powder is used to make cakes rise. When it's heated in the oven it thermally decomposes to produce a gas. ☐

Top Tips: Anything that takes heat in is **endothermic**. Endothermic reactions are not unusual in everyday life — think about what happens when you cook eggs and use baking powder.

Bond Energies

Q1 The **energy level diagrams** below represent the energy changes in five chemical reactions.

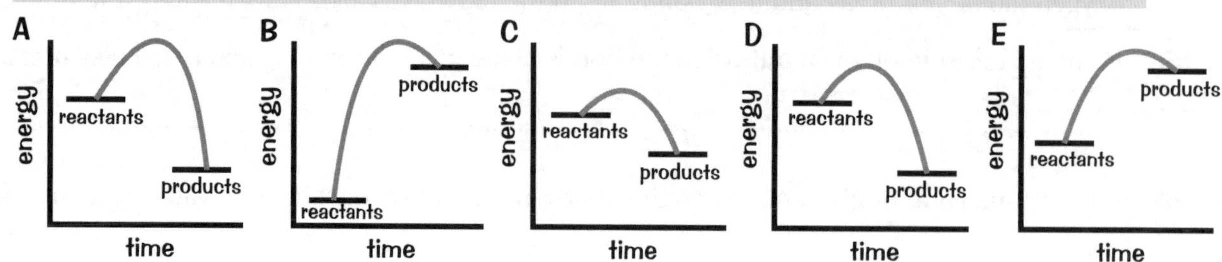

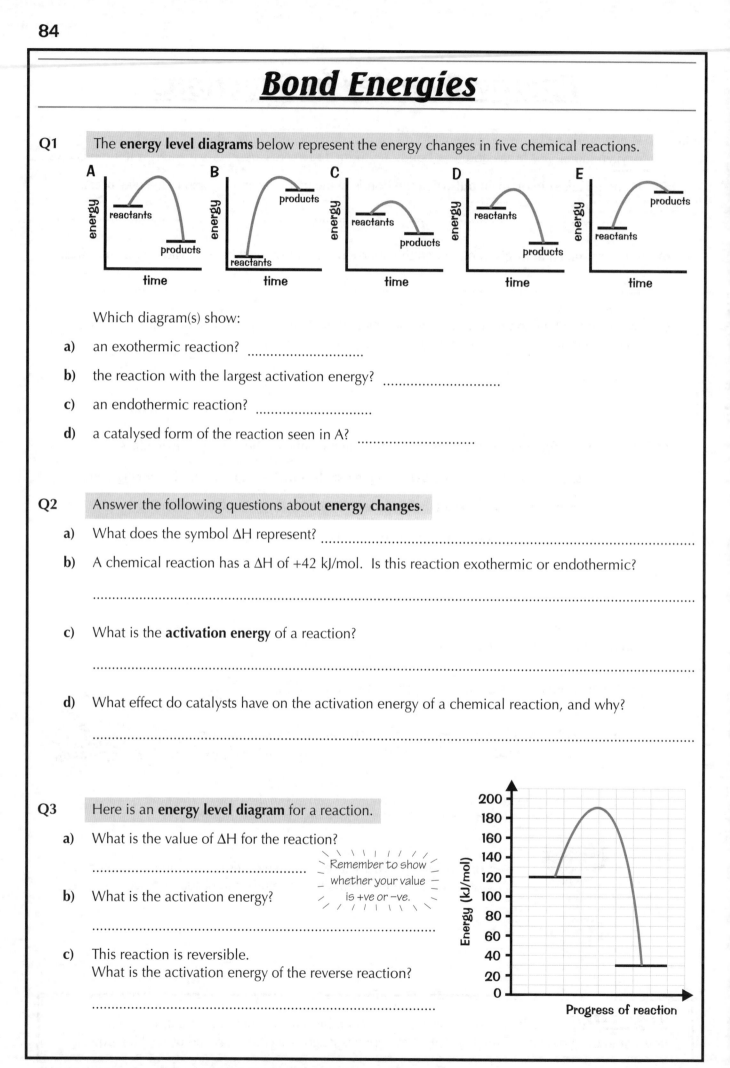

Which diagram(s) show:

a) an exothermic reaction? ....

b) the reaction with the largest activation energy? ....

c) an endothermic reaction? ....

d) a catalysed form of the reaction seen in A? ....

Q2 Answer the following questions about **energy changes**.

a) What does the symbol ΔH represent? ....

b) A chemical reaction has a ΔH of +42 kJ/mol. Is this reaction exothermic or endothermic?

....

c) What is the **activation energy** of a reaction?

....

d) What effect do catalysts have on the activation energy of a chemical reaction, and why?

....

Q3 Here is an **energy level diagram** for a reaction.

a) What is the value of ΔH for the reaction?

....

Remember to show whether your value is +ve or −ve.

b) What is the activation energy?

....

c) This reaction is reversible.
What is the activation energy of the reverse reaction?

....

Energy (kJ/mol)
200
180
160
140
120
100
80
60
40
20
0
Progress of reaction

Bond Energies

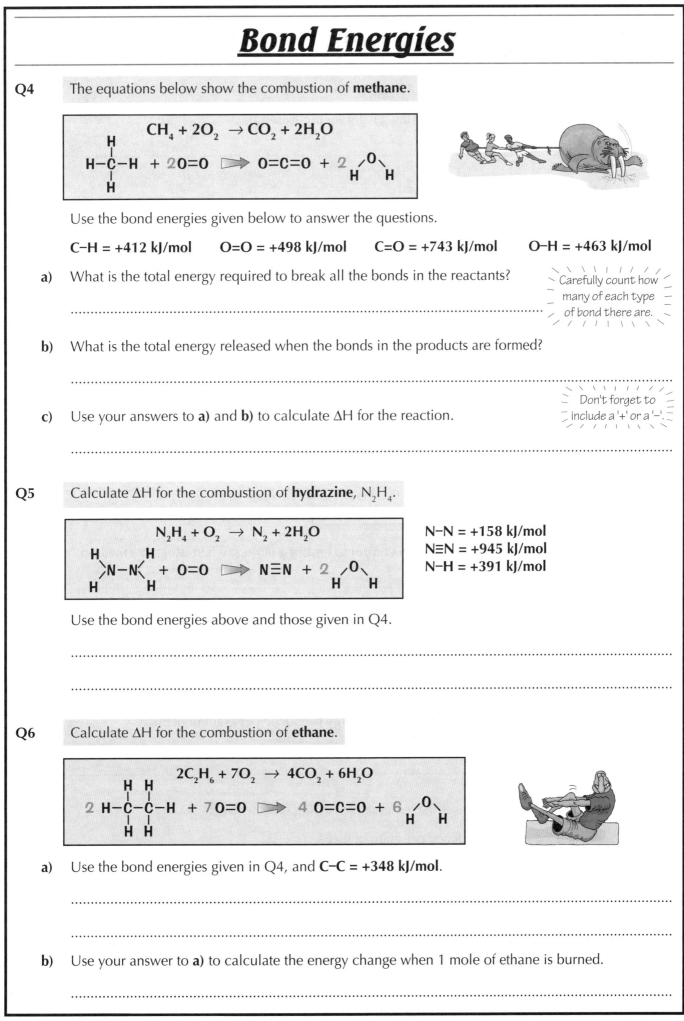

Q4 The equations below show the combustion of **methane**.

$$CH_4 + 2O_2 \rightarrow CO_2 + 2H_2O$$

H–C(H)(H)–H + 2 O=O ⟶ O=C=O + 2 H–O–H

Use the bond energies given below to answer the questions.

C–H = +412 kJ/mol **O=O = +498 kJ/mol** **C=O = +743 kJ/mol** **O–H = +463 kJ/mol**

a) What is the total energy required to break all the bonds in the reactants?

Carefully count how many of each type of bond there are.

....

b) What is the total energy released when the bonds in the products are formed?

....

c) Use your answers to **a)** and **b)** to calculate ΔH for the reaction.

Don't forget to include a '+' or a '–'.

....

Q5 Calculate ΔH for the combustion of **hydrazine**, N_2H_4.

$$N_2H_4 + O_2 \rightarrow N_2 + 2H_2O$$

(H)(H)N–N(H)(H) + O=O ⟶ N≡N + 2 H–O–H

N–N = +158 kJ/mol
N≡N = +945 kJ/mol
N–H = +391 kJ/mol

Use the bond energies above and those given in Q4.

....

....

Q6 Calculate ΔH for the combustion of **ethane**.

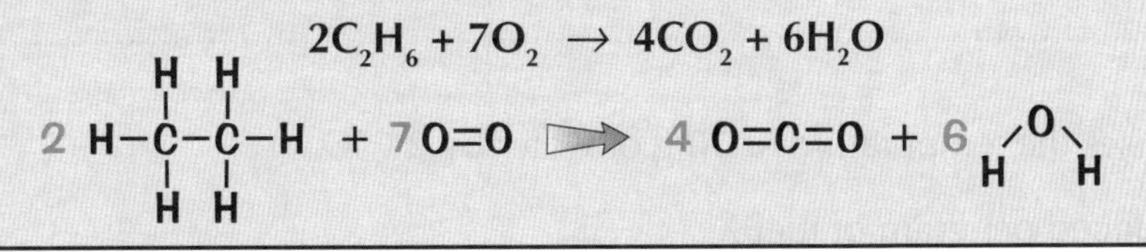

$$2C_2H_6 + 7O_2 \rightarrow 4CO_2 + 6H_2O$$

2 H–C(H)(H)–C(H)(H)–H + 7 O=O ⟶ 4 O=C=O + 6 H–O–H

a) Use the bond energies given in Q4, and **C–C = +348 kJ/mol**.

....

....

b) Use your answer to **a)** to calculate the energy change when 1 mole of ethane is burned.

....

Measuring the Energy Content of Fuels

Q1 Write down the formulas for calculating:

a) the energy transferred to water during a calorimetric experiment.

....

b) the energy output of a fuel per gram.

....

Q2 Ross wants to compare the **energy content** of two fuels, petrol and a petrol alternative, fuel X.

a) Draw a labelled diagram in the box to show the apparatus Ross could use for comparing the energy content of these two fuels in a simple calorimetric experiment.

b) Name **two variables** that Ross has to **control** to ensure a fair test when using this method.

....

c) He finds that **0.7 g** of petrol raises the temperature of **50 g** of water by **30.5 °C**.

i) Calculate the energy gained by the water.

....

ii) Use your answer to **i)** to calculate the energy produced per gram of petrol.
Give your answer in units of **kJ/g**.

....

d) Burning **0.8 g** of fuel X raises the temperature of **50 g** of water by **27 °C**.

Calculate the energy produced per gram of fuel X.

....

....

e) Using this evidence only, decide whether petrol or fuel X would make the better fuel.
Explain your choice.

....

Reversible Reactions

Q1 Use words from the list below to complete the following sentences about **reversible reactions**.

escape **reactants** **closed** **products** **react** **balance**

a) In a reversible reaction, the .... of the reaction can themselves .... to give the original ....

b) At equilibrium, the amounts of reactants and products reach a ....

c) To reach equilibrium the reaction must happen in a .... system, where products and reactants can't ....

Q2 Which of these statements about reversible reactions are **true** and which are **false**?

		True	False
a)	The position of an equilibrium depends on the reaction conditions.	☐	☐
b)	Upon reaching a dynamic equilibrium, the reactions stop taking place.	☐	☐
c)	You can move the position of equilibrium to get more product.	☐	☐
d)	At equilibrium there will always be equal quantities of products and reactants.	☐	☐
e)	Adding a catalyst moves the equilibrium position towards the products.	☐	☐

Q3 Substances A and B react to produce substances C and D in a **reversible reaction**.

$$2A_{(g)} + B_{(g)} \rightleftharpoons 2C_{(g)} + D_{(g)}$$

a) Give two reaction conditions which often affect the **position of equilibrium**.

1. .... 2. ....

b) The forward reaction is **exothermic**. Does the backward reaction give out or take in heat? Explain your answer.

....

....

c) If the temperature is raised, does the **forward** or **backward** reaction increase? ....

d) Explain why changing the temperature of a reversible reaction always affects the position of the equilibrium.

....

e) What effect will changing the **pressure** have on the position of equilibrium in this reaction? Explain your answer.

....

The Haber Process

Q1 The Haber process is used to make **ammonia**. The equation for the reaction is:

$$N_2(g) + 3H_2(g) \rightleftharpoons 2NH_3(g)$$

a) Name the reactants in the forward reaction.

....

b) Which side of the equation has more molecules?

....

c) How should the pressure be changed in order to produce more ammonia? Explain your answer.

....

....

Q2 The **industrial conditions** for the Haber process are carefully chosen.

a) What conditions are used? Tick one box.

☐ **1000 atmospheres, 450 °C**	☐ **200 atmospheres, 1000 °C**	☐ **450 atmospheres, 200 °C**	☐ **200 atmospheres, 450 °C**

b) Explain why this pressure is used.

....

....

Q3 In the Haber process reaction, the **forward** reaction is **exothermic**.

a) What effect will raising the temperature have on the **amount** of ammonia formed?

....

b) Explain why a high temperature is used industrially.

....

....

c) What happens to the leftover nitrogen and hydrogen?

....

Top Tips: Changing the conditions in a reversible reaction to get more product sounds great, but don't forget that these conditions might be too difficult or expensive for factories to produce, or they might give a reaction that's too slow to be profitable.

Gas Tests

Q1 Which piece(s) of apparatus shown could be used to **collect**:

A B C D

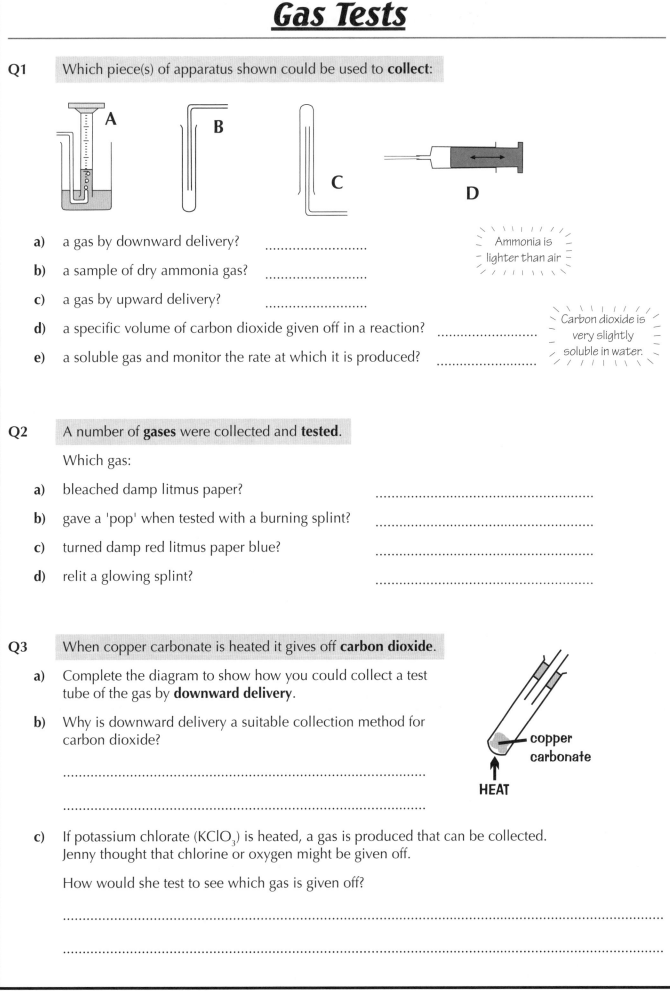

a) a gas by downward delivery? ....

b) a sample of dry ammonia gas? ....

c) a gas by upward delivery? ....

d) a specific volume of carbon dioxide given off in a reaction? ....

e) a soluble gas and monitor the rate at which it is produced? ....

Ammonia is lighter than air

Carbon dioxide is very slightly soluble in water.

Q2 A number of **gases** were collected and **tested**.

Which gas:

a) bleached damp litmus paper? ....

b) gave a 'pop' when tested with a burning splint? ....

c) turned damp red litmus paper blue? ....

d) relit a glowing splint? ....

Q3 When copper carbonate is heated it gives off **carbon dioxide**.

a) Complete the diagram to show how you could collect a test tube of the gas by **downward delivery**.

b) Why is downward delivery a suitable collection method for carbon dioxide?

....

....

copper carbonate

HEAT

c) If potassium chlorate ($KClO_3$) is heated, a gas is produced that can be collected. Jenny thought that chlorine or oxygen might be given off.

How would she test to see which gas is given off?

....

....

Tests for Positive Ions

Q1 Robert adds a solution of **sodium hydroxide** to a solution of **calcium chloride**. The formula of the calcium ion is Ca^{2+}.

a) What would Robert observe?

....

b) Write the balanced symbol equation for the reaction, including state symbols.

....

c) Write the balanced **ionic equation** for this reaction, including state symbols.

....

Q2 Les had four samples of **metal compounds**. He tested each one by placing a small amount on the end of a wire and putting it into a Bunsen flame. He observed the **colour of flame** produced.

a) Draw lines to match each of Les's observations to the metal cation producing the coloured flame.

brick-red flame	Na^{+}
yellow/orange flame	Cu^{2+}
blue-green flame	K^{+}
lilac flame	Ca^{2+}

b) Les wants to make a firework which will explode in his local football team's colour, **lilac**. Which of the following compounds should he use? Circle your answer.

silver nitrate **sodium chloride** **barium sulfate**

potassium nitrate **calcium carbonate**

Q3 Amy added a few drops of **sodium hydroxide** solution to solutions of various **metal compounds**.

a) Complete the balanced ionic equation for the reaction of iron(II) ions with hydroxide ions.

Fe^{2+}(........) + ... OH^{-}(aq) → .... (s)

b) Write a balanced ionic equation for the reaction of **iron(III) ions** with hydroxide ions.

Don't forget state symbols.

....

c) Amy added a few drops of sodium hydroxide solution to **aluminium sulfate solution**. She continued adding sodium hydroxide to excess. What would she observe at each stage?

....

....

Tests for Positive Ions

Q4 Claire was given a solid sample of an ionic compound. She was told that it was thought to be **ammonium chloride**.

a) Describe, in detail, how she would test for the presence of the **ammonium ion**.

....

....

b) What would she **observe** at each stage?

....

....

c) Write an **ionic equation** for the reaction that occurs.

....

Q5 Select compounds from the box to match the following statements.

KCl	**LiCl**	**$FeSO_4$**	**NH_4Cl**
$FeCl_3$	**$Al_2(SO_4)_3$**	**NaCl**	
$CuSO_4$	**$CaCl_2$**	**$MgCl_2$**	**$BaCl_2$**

a) This compound forms a blue precipitate with sodium hydroxide solution. ....

b) This compound forms a white precipitate with sodium hydroxide that dissolves if excess sodium hydroxide is added. ....

c) This compound forms a green precipitate with sodium hydroxide solution. ....

d) This compound forms a reddish brown precipitate with sodium hydroxide solution. ....

e) This compound reacts with sodium hydroxide to release a pungent gas. ....

f) This compound reacts with sodium hydroxide to form a white precipitate, and it also gives a brick-red flame in a flame test. ....

Top Tips: Right, this stuff needs to be learnt properly. Otherwise you'll be stuck in your exam staring at a question about the colour that some random solution goes when you add something you've never heard of before to it, and all you'll know is that ammonia smells of cat wee.

Tests for Negative Ions

Q1 Give the chemical formulas of the **negative ions** present in the following compounds.

a) barium sulfate .... **b)** potassium iodide .... **c)** silver bromide ....

Q2 Choose from the words given to complete the passage below.

carbon dioxide	limewater	hydrochloric acid	sodium hydroxide	hydrogen

A test for the presence of carbonates in an unidentified substance involves reacting it with dilute .... If carbonates are present then .... will be formed. You can test for this by bubbling it through .... to see if it becomes milky.

Q3 Answer the following questions on testing for **sulfate** and **sulfite** ions.

a) Which two **chemicals** are used to test for sulfate ions?

....

b) What would you **see** after adding these chemicals to a sulfate compound?

....

c) **i)** What substance is used to test for sulfite ions? ....

ii) Describe what you would see after adding this substance to a sulfite compound.

....

....

Q4 Deirdre wants to find out if a soluble compound contains **chloride**, **bromide** or **iodide** ions. Explain how she could do this.

....

....

....

Q5 Complete the following equations for **tests for negative ions**.

a) $Ag^+(aq)$ + ... → $AgCl(s)$

b) $2HCl(aq) + Na_2CO_3(s) \rightarrow 2NaCl(aq)$ + ...(l) + ...(g)

c) .... + ... → $BaSO_4(s)$

No...
You're being a bit negative today, aren't you?

Tests for Acids and Alkalis

Q1 Acids and alkalis can be tested for using **indicators**.

a) Complete the following statement about litmus indicator with the correct colours.

Acids turn .... litmus ...., and alkalis turn .... litmus ....

b) Which ions are always present in an acid? ....

c) How would you test for the presence of an acid other than using an indicator?
Describe the result of this test if an acid is present.

....

....

....

d) Which ions are always present in an alkali? ....

e) Other than using an indicator, how would you test for the presence of an alkali?
Describe the result of this test if an alkali is present.

....

....

....

Q2 Ammonia gas can be prepared in the laboratory by heating solid ammonium chloride with solid **calcium hydroxide**.

a) Write a balanced **symbol equation** for this preparation of ammonia.

....

b) Write the **ionic equation** for this reaction.

....

c) Describe how you could test for ammonia.

....

....

d) What colour will a solution of calcium hydroxide be with **phenolphthalein**?

....

Top Tips: Acids and alkalis is important stuff — and being able to explain how you'd identify H^+ and OH^- ions is a big indicator (ha ha) to the examiners of how much you know. So get learning...

Tests for Organic Compounds

Q1 Answer the following questions on heating **organic compounds**.

a) What colour is the flame when organic compounds are burnt in air?

....

b) Gus ignites two samples of hydrocarbons. One is propane, C_3H_8, and the other is octane, C_8H_{18}. Which sample will burn with a smokier flame? Explain why.

....

....

Q2 Pauline investigates the properties of two liquid organic compounds. She adds 1 cm^3 of **bromine water** to a 5 cm^3 sample of each compound. She then shakes each sample for 10 seconds and records her observations.

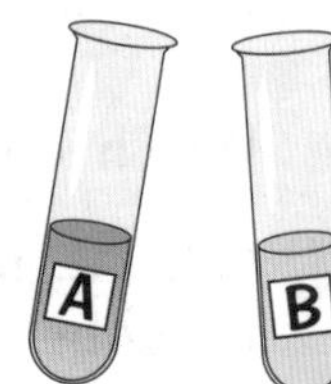

a) **i)** Give three examples of controlled variables in Pauline's investigation.

1. ....

2. ....

3. ....

ii) Explain why she controlled these variables.

....

Pauline presents her results in the table shown.

Organic compound	Colour of bromine water after shaking
A	colourless
B	orange

b) Which compound is a **saturated** hydrocarbon? ....

c) Which compound could be an **alkene**? ....

d) Which of the following could be the structural formula of organic compound **A**? Circle all possibilities.

```
  H H H         H H H H H         H     H H         H     H H
  | | |         | | | | |         |     | |         |     | |
H-C-C-C-H     H-C-C-C-C-C-H     H-C-C=C-C-C-H     H-C-C≡C-C-C-H
  | | |         | | | | |         | | | | |         |     | |
  H H H         H H H H H         H H H H H         H     H H
```

Tests for Organic Compounds

Q3 A sample of a hydrocarbon is burnt completely in air.
8.8 g of **carbon dioxide** and **5.4 g** of **water** are produced.

Multiply the mass of CO_2 produced by the proportion of carbon in CO_2.

a) Calculate the mass of **carbon** in the carbon dioxide.

....

....

b) Calculate the mass of **hydrogen** in the water.

....

....

c) Calculate the number of **moles** of carbon and of hydrogen in the sample of this hydrocarbon.

No. of moles = mass ÷ relative atomic mass.

....

....

d) What is the **empirical formula** of this hydrocarbon?

Find the simplest ratio of moles of C : moles of H.

....

....

ERROR!

Q4 A sample of a hydrocarbon is burnt completely in air. **4.4 g** of carbon dioxide and **1.8 g** of water are formed. What is the **empirical formula** of the hydrocarbon?

....

....

....

....

Q5 An organic compound contains only carbon, hydrogen and oxygen.
0.8 g of the compound is burnt completely in air. **1.1 g** of carbon dioxide and **0.9 g** of water are formed. What is the compound's **empirical formula**?

Watch out — some of the oxygen in the products came from the air, NOT the organic compound. You need to subtract the masses of C and H from the compound's mass to find the mass of O.

....

....

....

....

Instrumental Methods

Q1 Forensic scientists use **instrumental methods** to analyse substances found at crime scenes.

a) Suspects in criminal cases can only be held for a short period of time without being charged. In light of this, why are instrumental methods useful for preparing forensic evidence?

....

b) Give **two** other advantages of using instrumental methods.

....

Q2 **Atomic absorption spectroscopy** is used to identify metals. Ian compares the **absorption spectrum** of an unknown element to a set of absorption spectra from known elements.

unknown element | cadmium, Cd | lithium, Li | sodium, Na

a) Identify the **unknown** element. ....

b) Give **one** industry in which this method is used. ....

Q3 Answer the following questions on **mass spectrometry**.

a) A sample of a steel is vaporised and a mass spectrum taken. The spectrum identifies elements with relative atomic masses of 52, 55 and 56. What **elements** are present in the steel?

....

b) A mass spectrum shows that an alkene has a relative molecular mass of 70. What is the molecular formula of the alkene?

Alkenes have the general formula C_nH_{2n}.

....

c) A mass spectrum of a **pure** sample of the element antimony is taken. However two relative atomic masses are detected by the spectrometer, one at 121 and one at 123. Suggest why **two** relative atomic masses were detected.

....

Q4 **NMR spectroscopy** is a powerful instrumental method of analysis.

a) What do the letters **NMR** stand for? ....

b) What type of compound can this method analyse? ....

c) Which **atoms** in these compounds does NMR give you information about?

....

Water

Q1 The diagram below shows the **water cycle**.

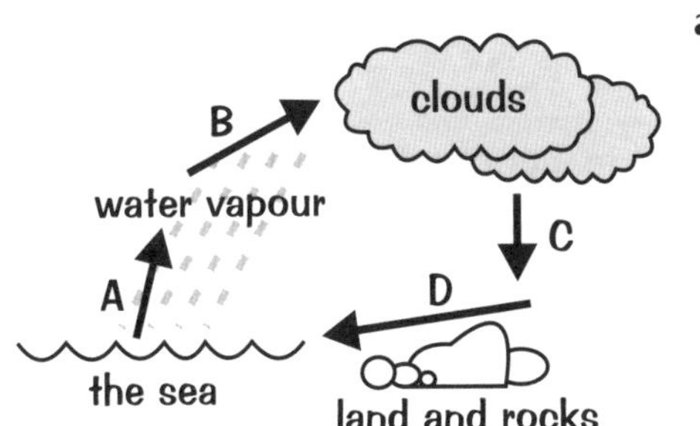

a) Write the correct letter (A, B, C or D) next to each label below to show where it belongs on the diagram.

air rises, water condenses ....

water flows ....

evaporation ....

rain ....

b) At which stage on the diagram is water separated from minerals in the sea? ....

c) At which stage do small amounts of minerals dissolve in water? ....

Q2 Water is sometimes known as the **universal solvent** because so many substances dissolve in it.

a) Tick the appropriate columns to show whether the compounds shown are **soluble** or **insoluble** in water.

SALT	SOLUBLE	INSOLUBLE
sodium sulfate		
ammonium chloride		
lead nitrate		
silver chloride		
lead sulfate		
potassium chloride		
barium sulfate		

b) Would you expect the following compounds to be **soluble** or **insoluble** in water? Justify your answers.

i) Potassium chromate. .... because ....

....

ii) Vanadium nitrate. .... because ....

....

Q3 Choose from the words in the box to fill in the blanks in the passage below.

power stations	alkali	fertilisers	sulfur	lead	households
rocks	distilled	acid	car washes	pure	impure

Rainwater is usually quite .... However, it may dissolve .... dioxide, which has been released from .... or car exhausts. This creates harmful .... rain. As rainwater flows over fields of crops it can dissolve chemicals from .... Minerals can also be dissolved as water passes over .... Water has to be treated before it can be supplied to ....

Water

Q4 Tick the boxes next to any of these statements that are **true**.

- [ ] All chloride salts are soluble in water.
- [ ] Water is essential for life because many biological reactions take place in solution.
- [ ] All nitrates are soluble in water, except for silver nitrate and lead nitrate.
- [ ] Fertilisers are often ammonium salts which are soluble in water.
- [ ] Many covalent compounds, like wax and petrol, don't dissolve in water.

Q5 When **sodium chloride** dissolves in water, the sodium and chloride ions, **Na^+** and **Cl^-**, are separated by the water molecules.

a) Water can dissolve ionic compounds because of the slight charges on either side of the water molecule. On the diagram below, label the atoms in the water molecule with their chemical symbols and show where the slight negative charge(s) and slight positive charge(s) are found.

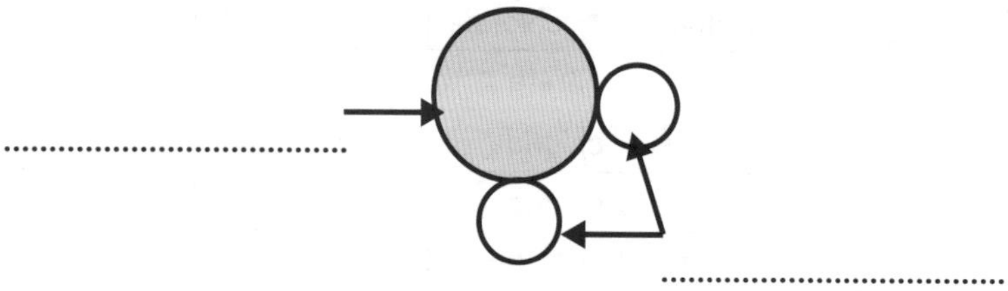

....

....

b) The diagram below is intended to show how water molecules interact with sodium and chloride ions when they are dissolved.

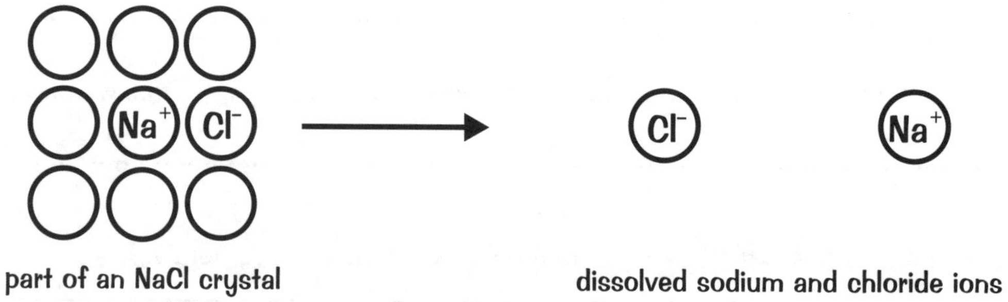

Na^+ Cl^- → Cl^- Na^+

part of an NaCl crystal

dissolved sodium and chloride ions

i) Label the remaining ions in the crystal.

ii) What holds the ions in the crystal together?

....

iii) Draw **four** water molecules around **each** dissolved ion above.

Top Tips: Go back to the basics when thinking about how ions interact. Opposite charges attract (no matter how small), so a positive ion attracts a negative ion and vice versa.

Hard Water

Q1 Tick the boxes to show whether the following sentences are **true** or **false**.

		True	False
a)	Rainwater which passes over limestone and chalk rocks becomes hard.	☐	☐
b)	Water can be softened by removing chloride and carbonate ions from the water.	☐	☐
c)	Adding sodium chloride is one way of removing hardness from water.	☐	☐
d)	Scale is formed when soap is used with hard water.	☐	☐
e)	You can remove the hardness from water by adding sodium carbonate.	☐	☐

Q2 In an experiment to investigate the **causes** of **hardness** in water, soap solution was added to different solutions. 'Five-drop portions' were added until a sustainable lather was formed.

Solution	Drops of soap solution needed to produce a lather	Observations	Drops of detergent solution needed to produce a lather
distilled water	5	no scum	5
magnesium sulfate solution	35	scum formed	5
calcium chloride solution	30	scum formed	5
sodium chloride solution	5	no scum	5

a) Why must all the solutions be prepared from distilled water rather than tap water?

....

b) i) Which compounds caused hardness in the water?

....

ii) Explain how you know. ....

....

c) What role did the test using distilled water play in the experiment?

....

d) What is the advantage of using detergent solution rather than soap for washing?

....

Q3 Hard water can cause the build-up of **scale** in pipes, boilers and kettles.

a) What is the disadvantage of scale building up in a kettle?

....

b) What is the disadvantage of scale building up in pipes?

....

Hard Water

Q4 Explain how hard water becomes soft when it is passed through an **ion exchange resin**. Write an equation which includes **$Na_2Resin(s)$** as one of the reactants to help you.

$Na_2Resin(s)$ + ... → .... + ...

....

....

Q5 A teacher wanted to demonstrate how chalk (composed of $CaCO_3$) dissolves in rainwater to produce **hard water**, and how it forms **scale** when it is boiled. She carried out the following experiments.

a) A spatula measure of powdered calcium carbonate was added to some distilled water and stirred. Why didn't the water become hard?

....

b) Carbon dioxide was bubbled through the mixture of calcium carbonate and distilled water.

i) Complete the equation below to show the reaction that took place.

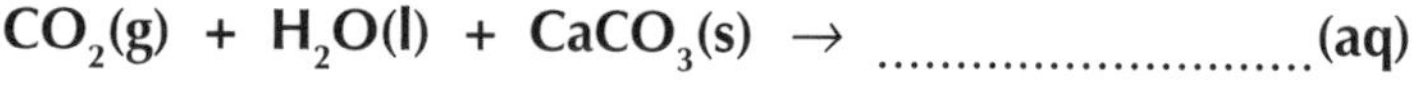

$CO_2(g) + H_2O(l) + CaCO_3(s) \rightarrow$ ....**(aq)**

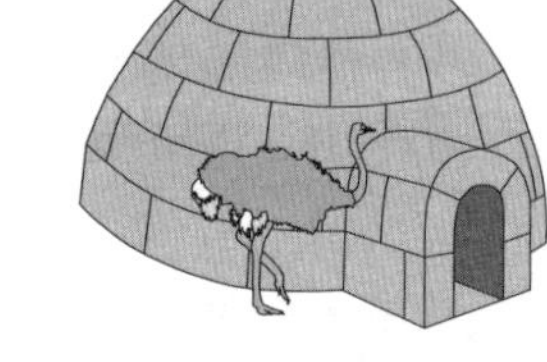

ii) Why did the water become hard?

....

....

c) A solution of calcium hydrogencarbonate was boiled in a beaker. As it boiled, a white precipitate formed.

i) Name the white precipitate formed.

....

ii) Write a full word equation for the reaction that happened.

....

d) Suggest one way that the white precipitate could be removed.

....

Top Tips: Hard water isn't very exciting, but at least it's not, well, hard. The only bits that will take some learning are the equations, especially that rather nasty calcium hydrogencarbonate one.

Water Quality

Q1 Choose the correct option from those given. Underline your answer.

a) Which type of water is most pure?

tap water river water distilled water sea water

b) Tap water can be passed through carbon filters. This removes:

ions which cause hardness chlorine taste microorganisms excess acidity

Q2 Draw lines to match the **water treatment processes** to the **substances removed**.

Water Treatment Process	Substance(s) removed
filtration	harmful microorganisms
chlorination	phosphates
adding iron compounds	acidity
adding lime (calcium hydroxide)	solids

Q3 The processes A to D are used in **water purification**.

A — filtration **B — distillation** **C — ion exchange** **D — boiling**

Which of these processes:

a) involves boiling and condensation? ....

b) kills microorganisms? ....

c) can be carried out using a gravel bed? ....

d) is used to soften water? ....

Q4 In 1995 it was estimated that **1 billion** people did **not** have access to clean drinking water.

a) Explain why so many people in developing countries don't have access to clean water.

....

....

b) Historically, how has life expectancy been linked to the ability to supply clean water?

....

Q5 One part of water treatment involves reducing **phosphate** and **nitrate** levels in drinking water.

a) Why are these substances removed? ....

b) What can be used to reduce nitrate levels in water? ....

c) **Phosphate** levels can be reduced by adding **iron compounds**. The phosphate is precipitated out as **iron phosphate**. How is the iron phosphate removed from the water?

....

Solubility

Q1 Look at the questions below and circle the best answer in each case.

a) The solubility of a solute is usually expressed in:

grams per 1000 grams of solvent — kilograms per 100 grams of solvent

moles per 100 grams of solvent — grams per 100 grams of solvent

b) A solution is saturated if:

the solution is colourless — no more solute dissolves at that temperature

the solution is clear — only stirring it causes more solute to dissolve

c) Gases dissolve more readily at **high** / **low** temperatures and at **high** / **low** pressures.

Q2 Choose from the words in the box to complete the passage below.

oxygen, fertiliser, aquatic, carbon dioxide, fizzy, bleach, underground, pressure, nitrogen, alcoholic, chlorine

Gas solubility is important for .... life. Fish need dissolved .... to survive. Small amounts of .... also dissolve in water, creating a solution which is used as a .... in the textile industry and to sterilise water supplies. .... dissolves well under .... and is used to make .... drinks.

Q3 The graph shows the solubility of **potassium nitrate** and **lead nitrate** at different **temperatures**.

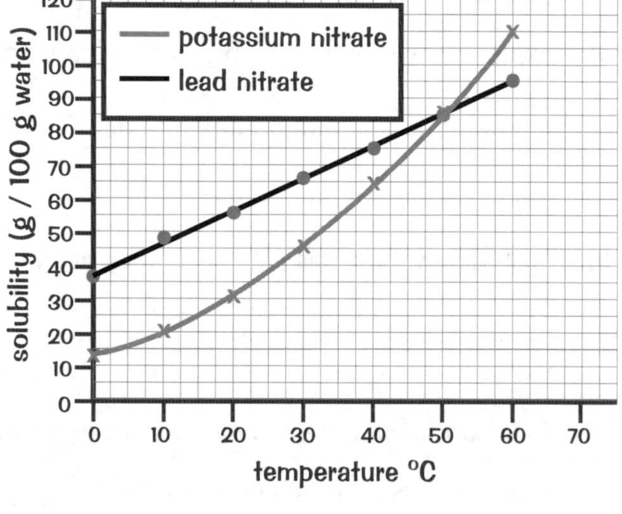

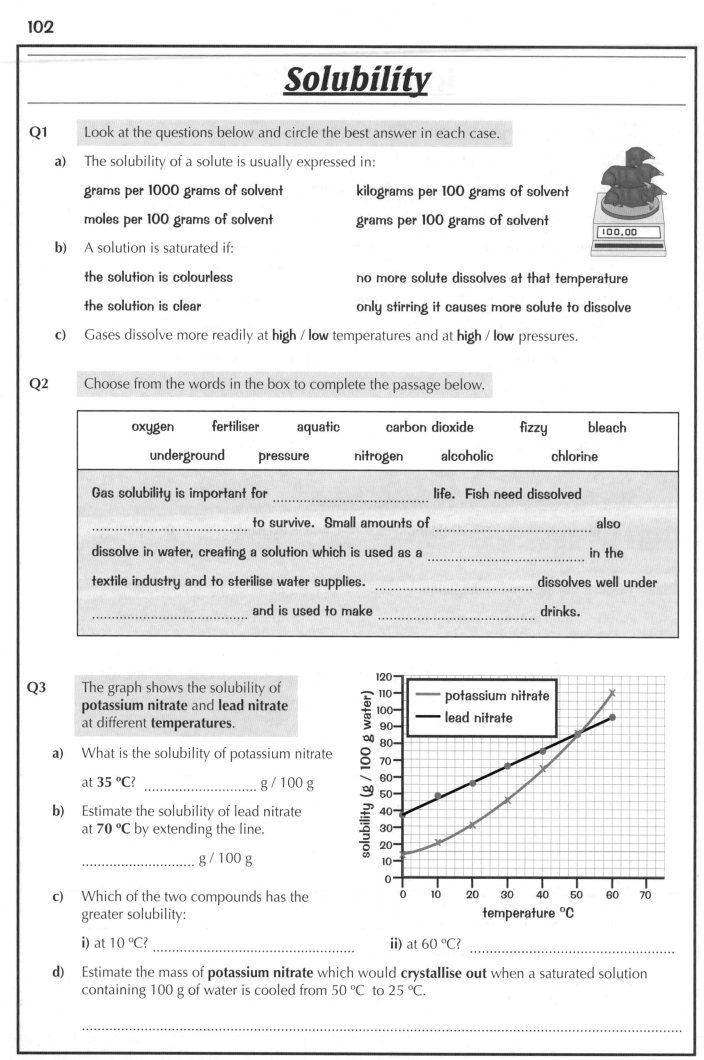

a) What is the solubility of potassium nitrate at **35 °C**? .... g / 100 g

b) Estimate the solubility of lead nitrate at **70 °C** by extending the line.

.... g / 100 g

c) Which of the two compounds has the greater solubility:

i) at 10 °C? ....

ii) at 60 °C? ....

d) Estimate the mass of **potassium nitrate** which would **crystallise out** when a saturated solution containing 100 g of water is cooled from 50 °C to 25 °C.

....

Detergents and Dry-Cleaning

Q1 The diagram shows a **detergent molecule**.

a) Complete the diagram by labelling the **hydrophilic** and **hydrophobic** sections of the molecule.

....

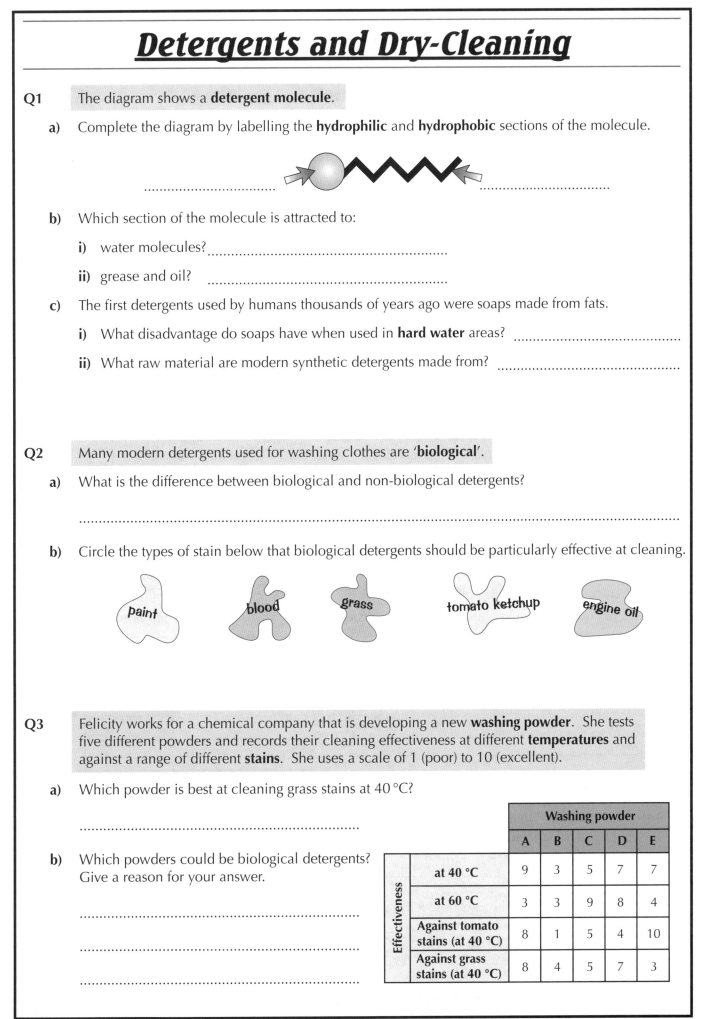

b) Which section of the molecule is attracted to:

i) water molecules? ....

ii) grease and oil? ....

c) The first detergents used by humans thousands of years ago were soaps made from fats.

i) What disadvantage do soaps have when used in **hard water** areas? ....

ii) What raw material are modern synthetic detergents made from? ....

Q2 Many modern detergents used for washing clothes are '**biological**'.

a) What is the difference between biological and non-biological detergents?

....

b) Circle the types of stain below that biological detergents should be particularly effective at cleaning.

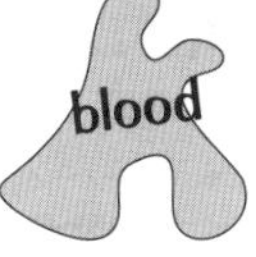

Q3 Felicity works for a chemical company that is developing a new **washing powder**. She tests five different powders and records their cleaning effectiveness at different **temperatures** and against a range of different **stains**. She uses a scale of 1 (poor) to 10 (excellent).

a) Which powder is best at cleaning grass stains at 40 °C?

....

b) Which powders could be biological detergents? Give a reason for your answer.

....

....

....

		Washing powder				
		A	**B**	**C**	**D**	**E**
Effectiveness	**at 40 °C**	9	3	5	7	7
	at 60 °C	3	3	9	8	4
	Against tomato stains (at 40 °C)	8	1	5	4	10
	Against grass stains (at 40 °C)	8	4	5	7	3

Detergents and Dry-Cleaning

Q4 Emily is trying to decide what **temperature** is best to wash her clothes at.

a) Why are **high** temperatures usually best for washing clothes?

....

b) Why shouldn't you wash clothes made from the following materials at high temperatures?

i) Wool ....

ii) Nylon ....

c) Her mum says, 'Washing clothes at high temperatures is environmentally unfriendly.' Explain why.

....

....

Q5 Use these words to complete the blanks.

ionic	intramolecular	solute	solvent	detergents
intermolecular	surrounded	covalent	solution	bonds

When a solid is dissolved in a liquid, ... forces form between the liquid molecules and the solid particles. These forces help to break .... between the solid particles, and the solid breaks up. A solid is dissolved when its particles are completely .... by liquid molecules. .... are used to help water dissolve substances. Some substances will not dissolve in water at all and another .... is required.

Q6 A chemical company is testing three new **solvents** for **dry-cleaning**.

	Solvent		
	A	B	C
Cost per 100 g (£)	0.40	0.15	0.20
Solubility of paint (g per 100 g of solvent)	12.1	0.1	10.3

a) What mass of solvent A is needed to dissolve 50 g of paint?

....

....

b) Which solvent would you expect to form the strongest intermolecular forces with paint molecules? Explain your answer.

....

c) Which solvent would you choose to buy if you were a buyer for a dry-cleaning company? Explain your choice.

....

....

Changing Equilibrium

Q1 Write the letter of one equation, A to C, to which the following statements apply.

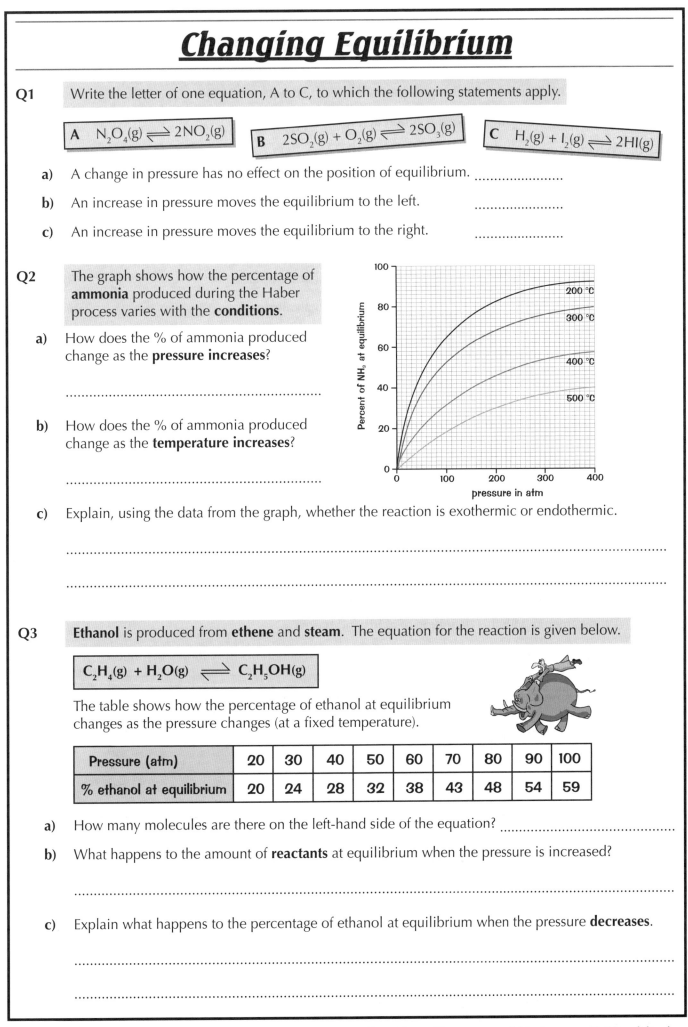

A $N_2O_4(g) \rightleftharpoons 2NO_2(g)$ **B** $2SO_2(g) + O_2(g) \rightleftharpoons 2SO_3(g)$ **C** $H_2(g) + I_2(g) \rightleftharpoons 2HI(g)$

a) A change in pressure has no effect on the position of equilibrium. ....

b) An increase in pressure moves the equilibrium to the left. ....

c) An increase in pressure moves the equilibrium to the right. ....

Q2 The graph shows how the percentage of **ammonia** produced during the Haber process varies with the **conditions.**

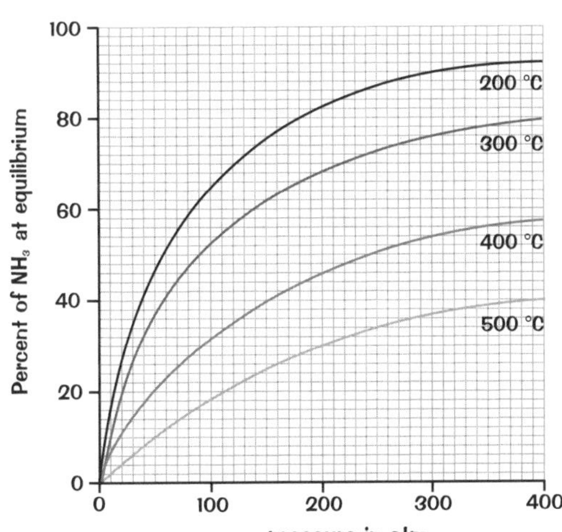

a) How does the % of ammonia produced change as the **pressure increases**?

....

b) How does the % of ammonia produced change as the **temperature increases**?

....

c) Explain, using the data from the graph, whether the reaction is exothermic or endothermic.

....

....

Q3 **Ethanol** is produced from **ethene** and **steam**. The equation for the reaction is given below.

$C_2H_4(g) + H_2O(g) \rightleftharpoons C_2H_5OH(g)$

The table shows how the percentage of ethanol at equilibrium changes as the pressure changes (at a fixed temperature).

Pressure (atm)	20	30	40	50	60	70	80	90	100
% ethanol at equilibrium	20	24	28	32	38	43	48	54	59

a) How many molecules are there on the left-hand side of the equation? ....

b) What happens to the amount of **reactants** at equilibrium when the pressure is increased?

....

c) Explain what happens to the percentage of ethanol at equilibrium when the pressure **decreases**.

....

....

Acid-Base Theories

Q1 Some particles found in acids and alkalis are: $H^+(aq)$ $H_2O(l)$ $OH^-(aq)$

a) Which particle would make a solution **acidic**? ....

b) Which particle would make a solution **alkaline**? ....

c) Which particle is a 'hydrated proton'? ....

Q2 **Arrhenius** studied acids and bases in the **1880s**. (You had to make your own entertainment in those days).

a) Complete the sentences below which outline his theory.

i) When mixed with ...., all acids release .... ions.

ii) When mixed with ...., all alkalis form .... ions.

b) It was known at the time that ammonia gas behaves as a **base**.
Explain why this prevented Arrhenius' ideas from being accepted at first.

....

c) Give one other reason why Arrhenius' theory was not immediately accepted.

....

Q3 One definition states that acid/base reactions involve **proton transfer**. This allows the reaction between **hydrogen chloride gas** and **ammonia gas** to be classified as an acid/base reaction.

$$NH_3(g) + HCl(g) \rightarrow NH_4^+Cl^-(s)$$

a) Write down the chemical symbol for a **proton**. ....

b) Who came up with this definition of acids and bases? ....

c) Complete the sentences to explain how the above reaction fits in with this definition.

When the two gases react, the hydrogen chloride behaves as .... by

....

At the same time the ammonia behaves as .... by

....

Q4 Explain how the following **bases** produce **alkaline** solutions.

a) potassium hydroxide, KOH ....

....

b) ammonia, NH_3 ....

....

Strong and Weak Acids

Q1 Which of the following are **strong acids**? Circle your answer(s).

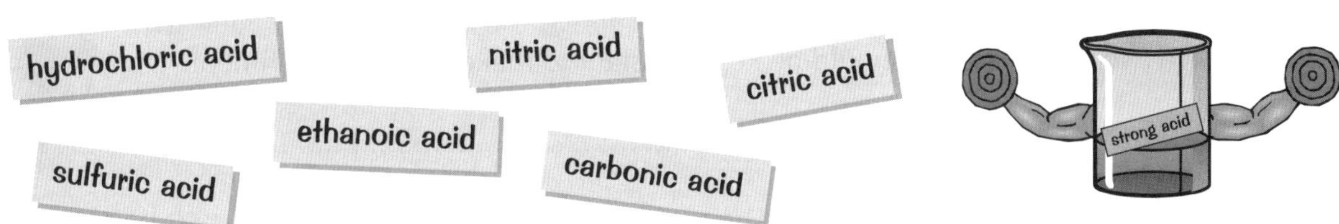

Q2 Tick the correct boxes to show whether the following statements are **true** or **false**.

		True	False
a)	Strong acids always have higher concentrations than weak acids.	☐	☐
b)	Strong acids ionise almost completely in water.	☐	☐
c)	Nitric acid ionises only very slightly in water.	☐	☐
d)	Weak acids ionise irreversibly in water.	☐	☐

Q3 Strong and weak acids react with **reactive metals** and with **carbonates** in the same way.

a) Complete the sentences by circling the correct word(s) in each pair.

i) Hydrochloric acid and ethanoic acid react with magnesium to give **hydrogen** / **oxygen**.

ii) Hydrochloric acid and ethanoic acid react with calcium carbonate to give **carbon dioxide** / **carbon monoxide**.

b) **i)** Do strong acids react **faster** or **slower** than weak acids? ....

ii) Explain why this is. ....

....

....

Q4 The graph shows the results of a reaction between **0.1 mol** of **magnesium** and **50 cm³** of **0.2 M hydrochloric acid**. (The acid is the limiting reactant.)

For each of the following reactions, sketch a curve on the graph above. (The acid is still the limiting reactant in both reactions.)

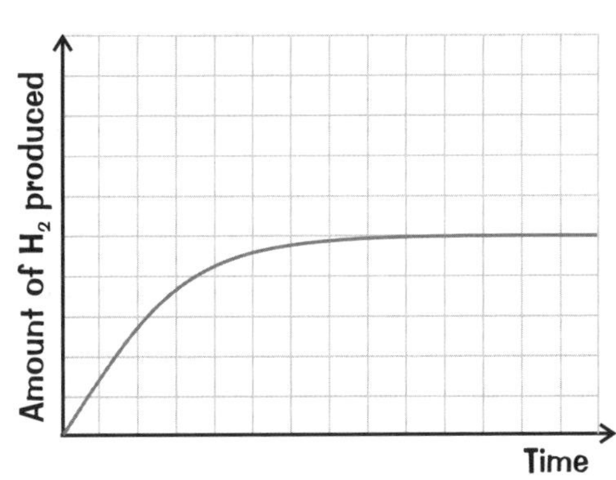

a) **Reaction 1** — 0.1 mol of magnesium is reacted with 50 cm^3 of 0.4 mol/dm^3 hydrochloric acid.

b) **Reaction 2** — 0.1 mol of magnesium is reacted with 50 cm^3 of 0.2 mol/dm^3 ethanoic acid.

Strong and Weak Acids

Q5 Fred has been asked to design an experiment to show that hydrochloric acid reacts faster than ethanoic acid with **magnesium ribbon**.

a) Give two variables that should be kept constant to make it a fair experiment.

1. ....

2. ....

b) Suggest how Fred could collect and measure the gas produced.

....

c) What difference would you expect to see in the amount of gas collected after:

i) both acids had reacted for 20 seconds?

....

ii) both acids had reacted completely?

....

Q6 Explain the following:

a) The pH of nitric acid is lower than the pH of lactic acid (of the same concentrations).

....

....

b) Hydrochloric acid is a better electrical conductor than carbonic acid (of the same concentration).

....

....

c) Weak acids and strong acids of the same concentration will produce the same amount of product.

....

....

d) Weak acids are used as descalers rather than strong acids.

....

....

Top Tips: Strong and weak acids are similar but different, and it's all down to how much they ionise. If you know that weak acids only ionise a bit, all the rest pretty much follows on from there.

Concentration

Q1 a) A solution contains **0.75 moles** of sulfuric acid in **1.5 dm³** of water. Calculate its **molar concentration**.

....

b) Calculate the **number of moles** of potassium manganate(VII) in **0.45 dm³** of a **4 M** solution of potassium manganate(VII).

....

c) A sample of **1.25 M** solution of sodium chloride contains **0.025 moles** of sodium chloride. Calculate its **volume** (in cm³).

....

d) **200 cm³** of a solution contains **0.25 moles** of iron hydroxide, $Fe(OH)_3$. Calculate its **molar concentration**.

....

e) What **volume** of a 1.6 M solution of calcium hydroxide contains **2 moles** of calcium hydroxide?

....

Q2 Convert the concentration of the following solutions from **mol/dm³** to **g/dm³**.

a) **2 mol/dm³** sodium hydroxide, NaOH. ....

....

b) **0.1 mol/dm³** glucose, $C_6H_{12}O_6$. ....

....

Q3 Convert the concentration of the following solutions from **g/dm³** to **mol/dm³**.

a) **5.6 g/dm³** potassium hydroxide, KOH. ....

....

b) **21 g/dm³** sodium hydrogencarbonate solution, $NaHCO_3$. ....

....

Q4 Barry dissolved **8 g** of copper(II) sulfate ($CuSO_4$) in **500 cm³** of water. Work out the concentration of this solution in **mol/dm³**.

....

....

....

Calculating Volumes

Q1 Choose from the following words to complete the passage.

atmosphere	volume	26	higher	mass	25	vole	mole	24

One .... of any gas will always occupy .... dm^3 when the .... is measured at a temperature of .... °C and a pressure of 1 .... If the volume is measured at a .... temperature, the molar volume of gas is increased.

Q2 The **limewater test** for carbon dioxide involves the reaction between carbon dioxide and calcium hydroxide, which is shown in the following equation:

$$CO_2 + Ca(OH)_2 \rightarrow CaCO_3 + H_2O$$

A solution of limewater containing 0.37 g of calcium hydroxide reacts with carbon dioxide at RTP.

a) What mass of **carbon dioxide** is needed to react completely with the limewater?

....

b) What **volume** does this amount of carbon dioxide occupy at RTP?

....

Q3 **Methane** burns in oxygen to produce carbon dioxide and water.

3.2 g of methane was completely burned in oxygen and the volume of each gas was measured. This was carried out at 112 °C and 1 atmosphere pressure. (1 mole of gas occupies 31 dm^3 at 112 °C and 1 atmosphere pressure.)

a) Write a balanced symbol **equation** for the reaction. Include state symbols.

....

b) What volume of **methane** was used in the reaction?

....

c) How much **oxygen** (in dm^3) reacted with the methane?

....

d) Calculate the **total volume of products** at 112 °C formed in this reaction.

....

....

Titrations

Q1 Label the following pieces of apparatus used in a **titration experiment**.

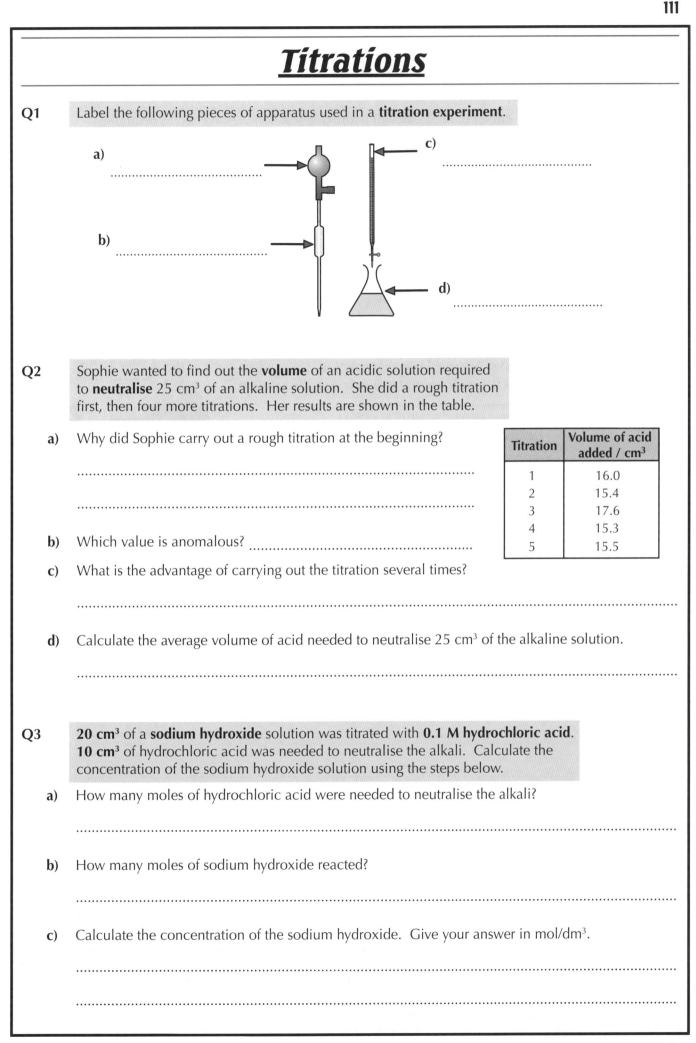

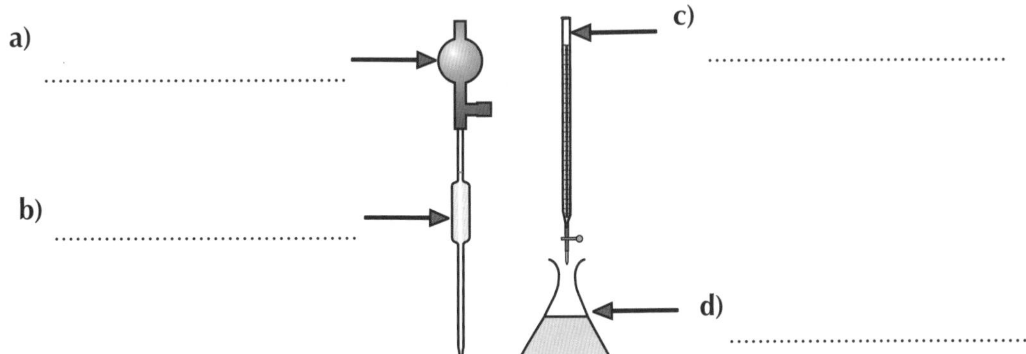

Q2 Sophie wanted to find out the **volume** of an acidic solution required to **neutralise** 25 cm³ of an alkaline solution. She did a rough titration first, then four more titrations. Her results are shown in the table.

Titration	Volume of acid added / cm³
1	16.0
2	15.4
3	17.6
4	15.3
5	15.5

a) Why did Sophie carry out a rough titration at the beginning?

b) Which value is anomalous?

c) What is the advantage of carrying out the titration several times?

d) Calculate the average volume of acid needed to neutralise 25 cm³ of the alkaline solution.

Q3 **20 cm³** of a **sodium hydroxide** solution was titrated with **0.1 M hydrochloric acid**. **10 cm³** of hydrochloric acid was needed to neutralise the alkali. Calculate the concentration of the sodium hydroxide solution using the steps below.

a) How many moles of hydrochloric acid were needed to neutralise the alkali?

b) How many moles of sodium hydroxide reacted?

c) Calculate the concentration of the sodium hydroxide. Give your answer in mol/dm³.

Titrations

Q4 In a titration, **12.5 cm³ of 0.04 M calcium hydroxide** solution was needed to neutralise **25 cm³** of **sulfuric acid**. Calculate the **concentration** of the sulfuric acid in mol/dm³.

$$H_2SO_4 + Ca(OH)_2 \rightarrow CaSO_4 + 2H_2O$$

....

....

....

Q5 In a titration, **10 cm³** of **hydrochloric acid solution** was used to neutralise **30 cm³** of **0.1 mol/dm³ potassium hydroxide solution.**

$$HCl + KOH \rightarrow KCl + H_2O$$

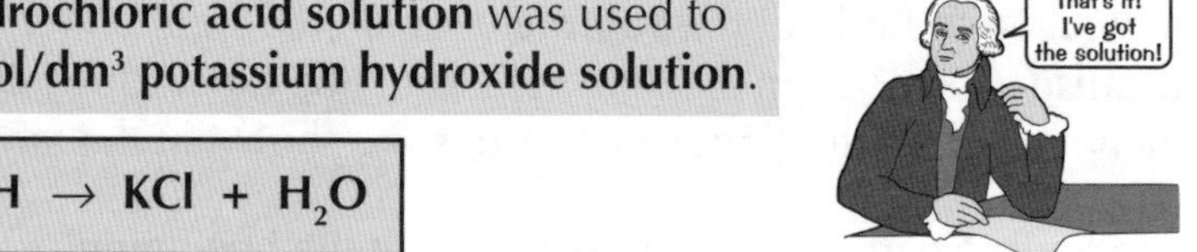

What was the concentration of the hydrochloric acid in moles per dm³?

....

....

....

END POINT

Q6 Brenda wants to find out the concentration of a solution of sodium hydroxide. She carries out a **titration** using **15 cm³** of **0.2 M** hydrochloric acid, and finds that it takes **22 cm³** of sodium hydroxide to neutralise the acid.

a) Calculate the concentration of the sodium hydroxide solution, in **mol/dm³**.

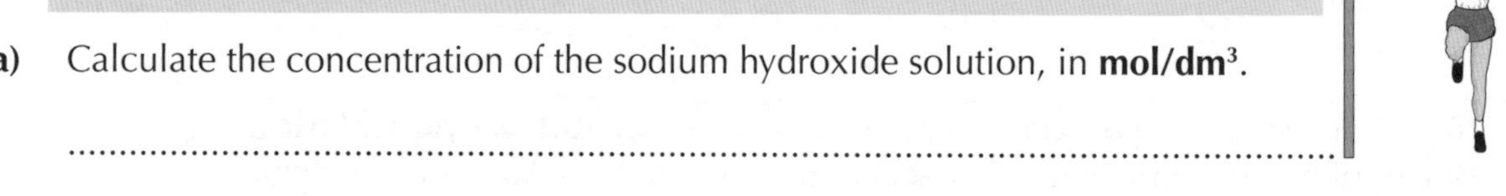

....

....

b) Calculate the concentration of the sodium hydroxide solution, in **g/dm³**.

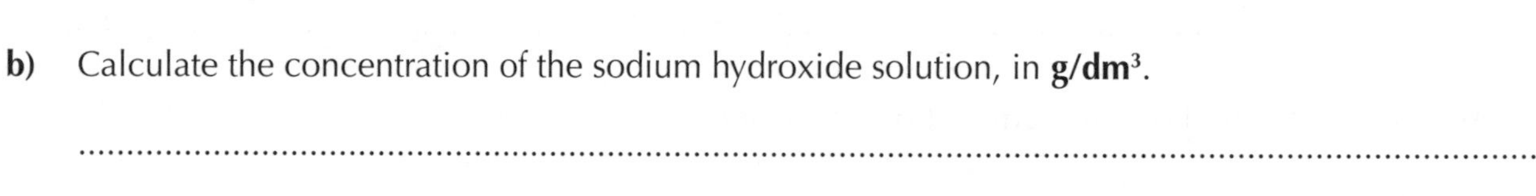

....

....

Top Tips: Aargh, calculations. As if Chemistry wasn't tricky enough without maths getting involved too (but at least it's not as bad as Physics). Actually, these aren't the worst calculations as long as you tackle them in stages and know your equations.

Electrolysis

Q1 Tick the correct boxes to show whether the following statements about the electrolysis of **molten sodium chloride** are **true** or **false**.

		True	False
a)	The chloride ions are oxidised.	☐	☐
b)	Chloride ions are attracted to the negative cathode.	☐	☐
c)	The sodium ions are reduced.	☐	☐
d)	Sodium ions are attracted to the positive anode.	☐	☐
e)	Electrolysis always involves either reduction or oxidation, never both.	☐	☐

Q2 Electroplating could be used to put a thin coat of **silver** onto a **nickel** fork.

a) Complete the diagram by labelling the **cathode** and **anode**.

b) What ion must the electrolyte contain?

....

pure silver strip

Q3 **Molten copper(II) chloride** is electrolysed using carbon electrodes.

a) Write the half-equation for the reaction at the **anode**. ....

b) Write the half-equation for the reaction at the **cathode**. ....

c) Write the full **ionic equation** for the electrolysis of copper(II) chloride.

....

Q4 Study the reactivity series and the table showing the products at the cathodes when different solutions of **ionic compounds** are electrolysed.

Ionic Compound Solution	Product at Cathode
sodium nitrate	hydrogen
copper sulfate	copper
sodium iodide	hydrogen
potassium chloride	hydrogen
silver nitrate	silver

reactivity (increasing upwards):
potassium
sodium
calcium
carbon
zinc
iron
lead
hydrogen
copper
silver

What do you notice about the substance released at the cathode and where it's found in the reactivity series?

....

....

Electrolysis

Q5 Roy electrolyses **potassium sulfate solution**, K_2SO_4.

a) Give the formulas of the **four** ions present in this solution. ....

b) i) Which ion is discharged at the anode? Explain your answer.

....

ii) Write a balanced symbol equation for this reaction.

....

c) i) Which ion is discharged at the cathode? Explain your answer.

....

ii) Write a balanced symbol equation for this reaction.

....

d) Suggest why the electrodes are made from graphite for the electrolysis of potassium sulfate.

....

Q6 The table shows the **ions** present in solutions of some ionic compounds and the **products** formed when these solutions are electrolysed. Use this table to answer the following questions.

SOLUTION	IONS PRESENT	CATHODE	ANODE
calcium nitrate	Ca^{2+}, NO_3^-, H^+, OH^-	hydrogen	oxygen
copper chloride	Cu^{2+}, Cl^-, H^+, OH^-	copper	chlorine
sodium sulfate	Na^+, SO_4^{2-}, H^+, OH^-	hydrogen	oxygen
potassium bromide	K^+, Br^-, H^+, OH^-	hydrogen	bromine

a) Where do the H^+ and OH^- ions present in the solutions come from? ....

b) i) Give the formulas of two negative ions that are discharged **less** easily than the OH^- ion.

....

ii) Give the formulas of two negative ions that are discharged **more** easily than the OH^- ion.

....

c) i) Give the formula of a positive ion that is discharged more easily than a H^+ ion. ....

ii) Give the formulas of the positive ions that are discharged less easily than H^+ ions.

....

Top Tips: Working out what substance is produced at each electrode isn't easy — especially when it's a dilute solution with loads of H^+ and OH^- ions running about as well. Just remember that OH^- ions will lose electrons in preference to NO_3^- ions and SO_4^{2-} ions, and that oxygen will be produced.

Electrolysis — Calculating Masses

Q1 Complete the table to show the amounts required to produce **1 mole** of each metal from its ions.

Metal ion	No. of moles of electrons	No. of faradays	No. of coulombs
Ca^{2+}			
K^{+}			
Al^{3+}			

Q2 a) Give two ways you could **increase** the amount of a substance produced during electrolysis.

1. ....

2. ....

b) Write down the formula for calculating the amount of charge transferred.

....

c) Calculate the charge transferred when:

i) 2.5 A has flowed for 15 s. ....

ii) 0.1 A has flowed for 30 mins. ....

d) Calculate the time (in minutes) 6 A needs to flow for to pass 4320 coulombs.

....

Q3 Molten **silver nitrate** was electrolysed for **40 minutes** using a current of **0.2 amps**.

a) Write the half-equation for the reaction at the **negative electrode**.

....

b) How many **coulombs** of charge flowed during the electrolysis?

....

c) How many **faradays** is this? ....

d) How many **moles** of silver were deposited at the cathode? ....

e) Calculate the **mass** of silver deposited at the cathode.

....

Top Tips: These calculations are HARD! So well done for making it to the end of the section. You'd better go and have a lie-down in a darkened room for a while now to recover.

Chemical Production

Q1 Widely used chemicals are often produced by **continuous production**.

a) Circle the correct words to complete the following sentences.

Continuous production is often used for the **small-** / **large-**scale production of chemicals.
It's **highly automated** / **labour-intensive**, which means there are **low** / **high** labour costs.
Products of a **high** / **low** consistency can be produced with a **high** / **low** risk of contamination.

b) State two **disadvantages** of continuous production.

1. ....

2. ....

c) Why are **pharmaceutical drugs** usually manufactured using batch production?

....

Q2 Use the following words to complete the blanks.

yield sufficient optimum rate recycled lowest

.... conditions are chosen to give the .... production cost per kg of product. This may mean that the conditions used do not have the highest .... of reaction or the highest percentage .... of product. However, both the rate and the yield must be high enough to give a .... daily yield of product. A low percentage yield is acceptable if the starting materials can be .... and reacted again.

Q3 Explain how the following affect the **production costs** of making a new substance.

a) Catalysts ....

....

b) Recycling raw materials ....

....

c) Automation ....

....

d) High temperatures ....

....

Alcohols

Q1 Tick the correct boxes to show whether the following statements are **true** or **false**.

		True	False
a)	Ethanol is a clear, colourless liquid at room temperature.	☐	☐
b)	Methanol is a non-volatile alcohol.	☐	☐
c)	Propanol is miscible with water.	☐	☐
d)	Methanol is less toxic than ethanol.	☐	☐

Q2 The molecular formula for **ethanol** can be written as $\mathbf{C_2H_5OH}$ or as $\mathbf{C_2H_6O}$.

a) What is the functional group found in all alcohols? ....

b) Explain why it is better to write ethanol's formula as C_2H_5OH.

....

Q3 Ethanol can be **dehydrated** to make **ethene**.

a) Which of the following use ethene in their production? Circle the correct answer(s).

sulfuric acid ammonia polymers esters plastics

b) Write a symbol equation for the conversion of ethanol into ethene.

....

c) What could be used as a catalyst for this process?

....

Q4 **Ethanol** is commonly used as a **solvent**.

a) Which part of ethanol's structure allows it to dissolve substances like hydrocarbons, oils and fats?

....

b) Which part of ethanol's structure allows it to mix with water and dissolve ionic compounds?

....

c) Ethanol is used in such things as glues, varnishes, printing inks, paints, deodorants and aftershaves. Give **two** properties of ethanol that often make it a good choice as a solvent.

....

....

Hint — these products need to 'dry'.

Carboxylic Acids and Esters

Q1 Tick the correct boxes to show whether the following statements are **true** or **false**.

		True	False
a)	Carboxylic acids have the functional group –COOH.	☐	☐
b)	There are six carbon atoms in every molecule of propanoic acid.	☐	☐
c)	The longer the hydrocarbon chain, the less soluble a carboxylic acid is in water.	☐	☐

Q2 Match up the following carboxylic acids to the correct statement.

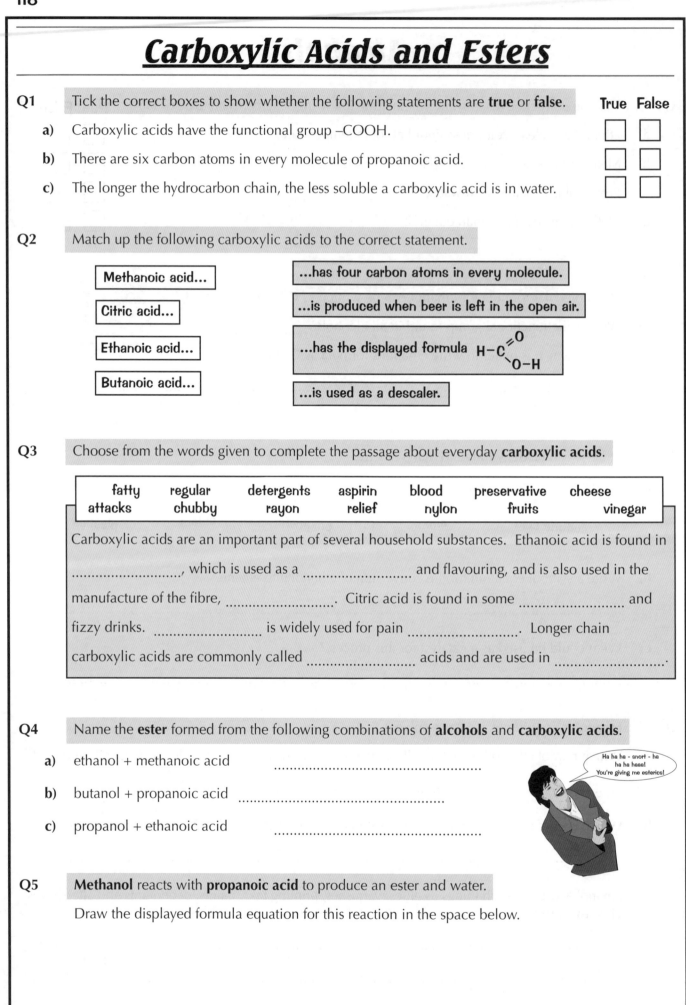

Methanoic acid...	...has four carbon atoms in every molecule.
Citric acid...	...is produced when beer is left in the open air.
Ethanoic acid...	...has the displayed formula H–C(=O)–O–H
Butanoic acid...	...is used as a descaler.

Q3 Choose from the words given to complete the passage about everyday **carboxylic acids**.

fatty regular detergents aspirin blood preservative cheese
attacks chubby rayon relief nylon fruits vinegar

Carboxylic acids are an important part of several household substances. Ethanoic acid is found in ...., which is used as a .... and flavouring, and is also used in the manufacture of the fibre, ... Citric acid is found in some .... and fizzy drinks. .... is widely used for pain .... Longer chain carboxylic acids are commonly called .... acids and are used in ....

Q4 Name the **ester** formed from the following combinations of **alcohols** and **carboxylic acids**.

a) ethanol + methanoic acid ....

b) butanol + propanoic acid ....

c) propanol + ethanoic acid ....

Ha ha ha - snort - ha ha ha haaa! You're giving me esterics!

Q5 **Methanol** reacts with **propanoic acid** to produce an ester and water.

Draw the displayed formula equation for this reaction in the space below.

Drug Development

Q1 Most modern drugs are made by a process called **staged synthesis**.

a) Describe how a drug **ABC** could be made by staged synthesis from compounds A, B and C.

....

....

b) How would a drug company make a 'family' of new drugs similar to ABC?

....

....

Q2 A drug company wants to make a **family** of compounds similar to a successful new drug, **PQR**. They have 2 **P-type** (P1 and P2), 2 **Q-type** (Q1 and Q2) and 2 **R-type** (R1 and R2) compounds.

a) List the possible combinations to show how they could make eight new compounds.

....

....

b) Calculate how many compounds the company could make if they had 15 of each type.

....

Q3 Compounds used in pharmaceutical drugs are often extracted from **plants**.

a) Describe the following steps in the extraction process.

A B C

Step A ....

Step B ....

Step C ....

b) New drugs may be tested on animals before being sold.

i) Give one argument for and one against testing new drugs on **animals**.

For ....

Against ....

ii) After animal testing, why are **human trials** of drugs also necessary?

....

Painkillers

Q1 Here are the **displayed formulas** of aspirin, paracetamol and ibuprofen.

Aspirin

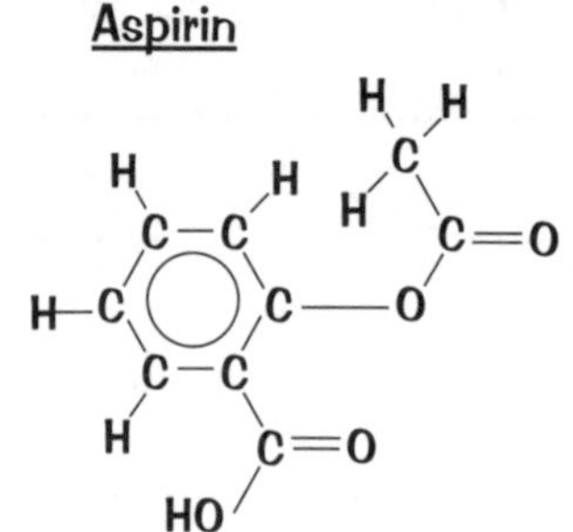

Paracetamol

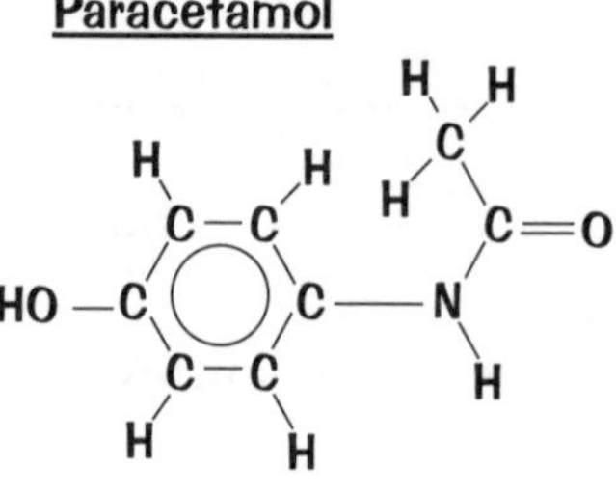

Ibuprofen

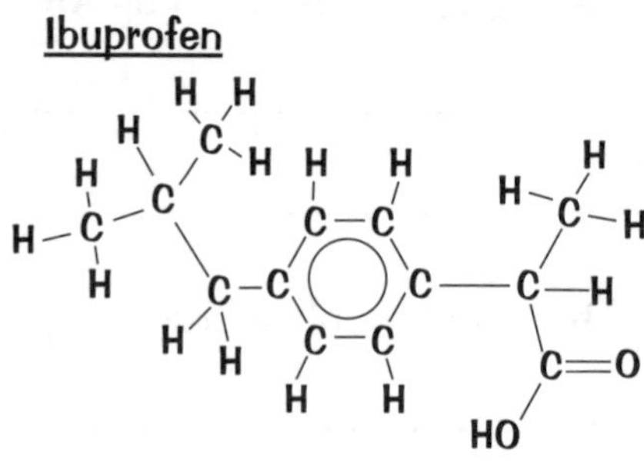

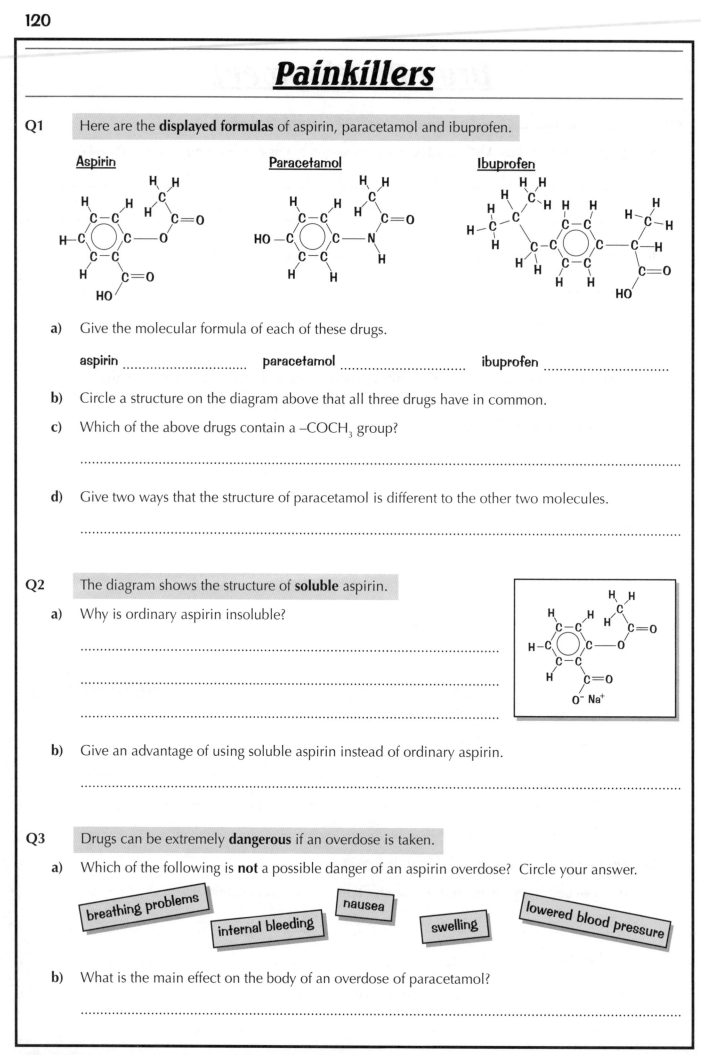

a) Give the molecular formula of each of these drugs.

aspirin .... **paracetamol** .... **ibuprofen** ....

b) Circle a structure on the diagram above that all three drugs have in common.

c) Which of the above drugs contain a $-COCH_3$ group?

....

d) Give two ways that the structure of paracetamol is different to the other two molecules.

....

Q2 The diagram shows the structure of **soluble** aspirin.

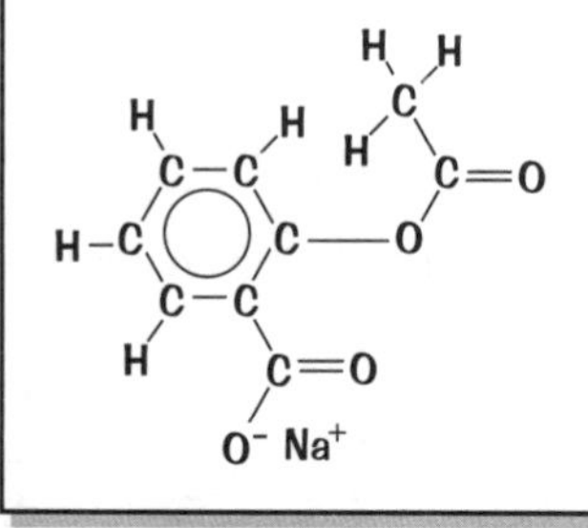

a) Why is ordinary aspirin insoluble?

....

....

....

b) Give an advantage of using soluble aspirin instead of ordinary aspirin.

....

Q3 Drugs can be extremely **dangerous** if an overdose is taken.

a) Which of the following is **not** a possible danger of an aspirin overdose? Circle your answer.

b) What is the main effect on the body of an overdose of paracetamol?

....

Making Sulfuric Acid

Q1 The Contact process is used to manufacture sulfuric acid.
Complete the table to show the **conditions** used in the Contact process.

Temperature:
Pressure:
Catalyst:

Q2 **Complete** and **balance** the following equations involved in the Contact process.

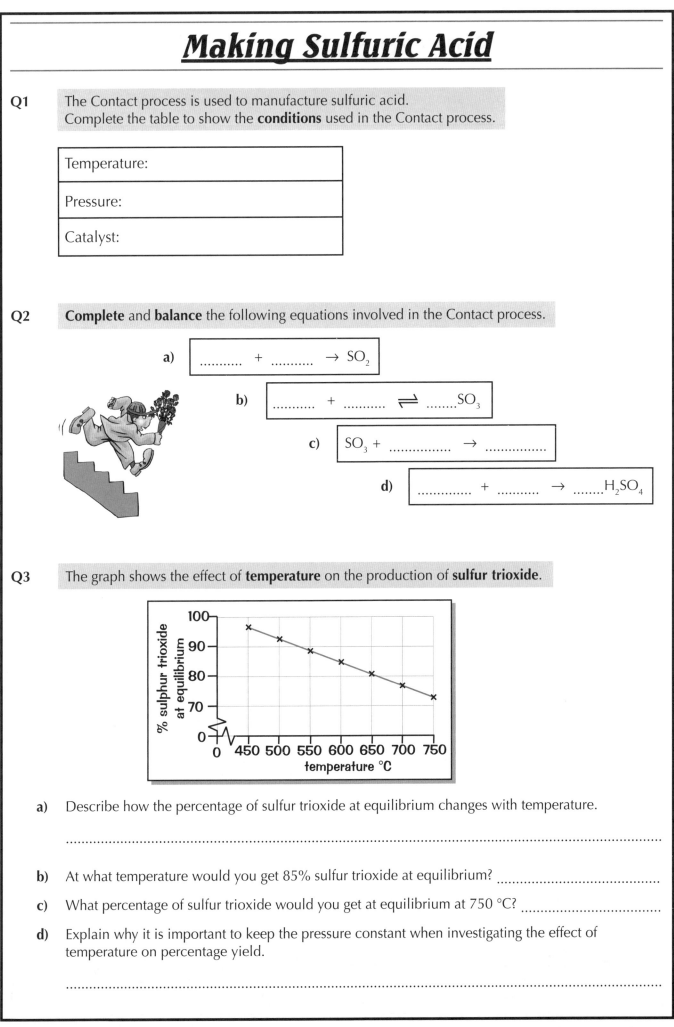

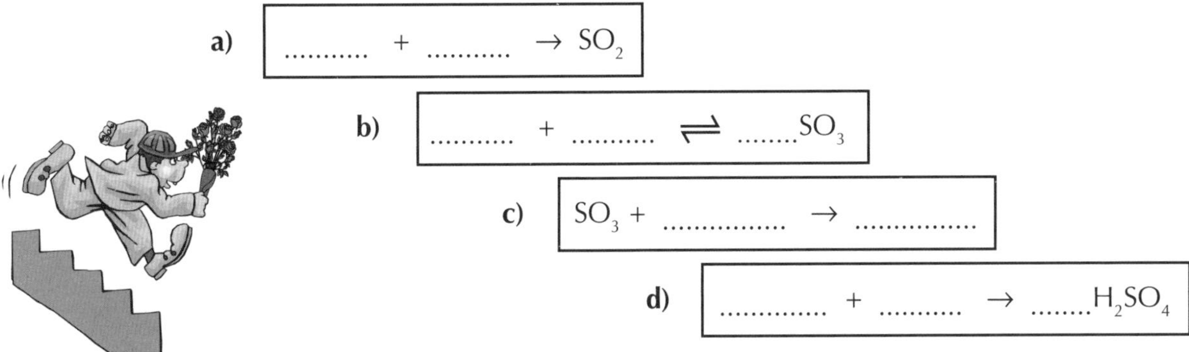

a) .... + ... → SO_2

b) .... + ... ⇌ ....SO_3

c) SO_3 + ... → ....

d) .... + ... → ....H_2SO_4

Q3 The graph shows the effect of **temperature** on the production of **sulfur trioxide**.

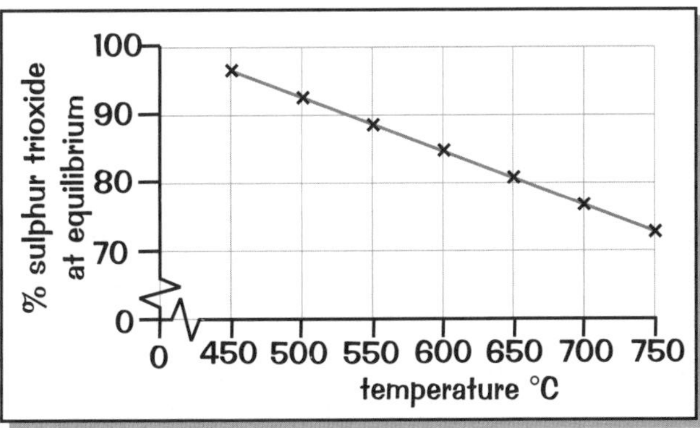

a) Describe how the percentage of sulfur trioxide at equilibrium changes with temperature.

....

b) At what temperature would you get 85% sulfur trioxide at equilibrium? ....

c) What percentage of sulfur trioxide would you get at equilibrium at 750 °C? ....

d) Explain why it is important to keep the pressure constant when investigating the effect of temperature on percentage yield.

....

Making Sulfuric Acid

Q4 Complete the following sentences by circling the correct word from each pair.

The **reduction / oxidation** of sulfur dioxide to sulfur trioxide is **exothermic / endothermic**.

When the temperature is increased, you get **more / less** sulfur trioxide.

If the temperature of any reaction is increased, the rate of the reaction **decreases / increases** because the particles have **more / less** energy.

A high temperature gives a **high / low** yield of sulfur trioxide, but produces it **slowly / quickly**.

Q5 As part of the Contact process, SO_3 is formed at a pressure of **1-2 atmospheres**.

a) Describe what would happen to the yield of SO_3 if the pressure was increased.

....

b) Explain why this is.

....

....

c) Give two reasons why this oxidation is not done at a high pressure.

1. ....

2. ....

Q6 There are **four main stages** involved in the Contact process. Write a number from 1 to 4 next to each of the descriptions below to show which stage it refers to.

a) Fuming sulfuric acid (oleum) is the product. ☐

b) The reaction is reversible. ☐

c) Concentrated sulfuric acid is formed. ☐

d) The product can be made by roasting sulfide ores. ☐

e) A catalyst is used. ☐

f) Neither oxygen nor water is a reactant. ☐

Top Tips: Sulfuric acid is a really important industrial chemical, it's used in things like fertiliser production, oil refining, iron and steel making and even for making explosives. It causes quite nasty burns though so workers have to be very careful when handling it.

Fuel Cells

Q1 Fill in the blanks to complete the passage below.

A fuel cell is an electrical cell that's supplied with a .... and .... and uses energy from the reaction between them to generate a ....

Q2 Give two **advantages** of using hydrogen fuel cells over alternative energy sources.

1. ....

2. ....

Q3 **Hydrogen** and **oxygen** react together in an **exothermic** reaction.
Sketch an energy level diagram in the box provided to show this reaction.

Q4 The diagram shows a hydrogen-oxygen **fuel cell**.

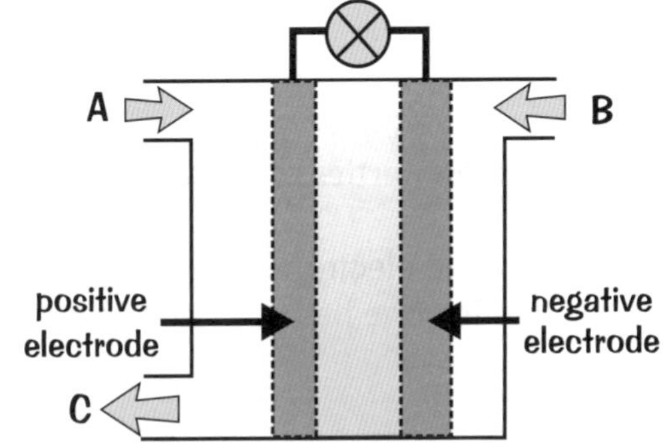

a) What goes into the cell at A and B?

A ....

B ....

b) What comes out of the cell at C?

....

c) What could the electrodes be made of? ....

d) What could be used as the electrolyte? ....

e) Write the symbol equation for the reaction that occurs at the:

i) negative electrode. ....

ii) positive electrode. ....

f) Write the overall equation for the reaction in the cell. ....

Transition Metals

Q1 Complete the passage below by circling the correct word(s) from each pair.

> Most metals are in the transition block found **at the left** / **in the middle** of the periodic table. The transition metals are usually **reactive** / **unreactive** with oxygen and water. They generally have high **densities** / **volatilities** and **low** / **high** melting points. They are **good** / **poor** conductors of heat and electricity. Their compounds are often **coloured** / **shiny** and, like the metals themselves, are effective **fuels** / **catalysts** in many reactions.

Q2 Transition metals and their compounds often make **good catalysts**.

Draw lines to match the metals and compounds below to the reactions they catalyse.

iron	converting natural oils into fats
manganese(IV) oxide	ammonia production
nickel	decomposition of hydrogen peroxide
vanadium pentoxide	sulfuric acid production

Q3 The **properties** of transition metals are due to the way their **electron shells** fill.

a) Give the electron arrangements of the following transition metals.
The first one has been done for you.

Titanium **2, 8, 10, 2**

i) Iron ....

ii) Vanadium ....

iii) Nickel ....

HINT: the atomic number is the same as the total number of electrons in an atom.

b) Transition metals often form more than one ion. Write down two different ions formed by:

i) Iron ....

ii) Copper ....

iii) Chromium ....

Transition Metals

Q4 'Chemical gardens' can be made by sprinkling **transition metal salts** into **sodium silicate solution**. Transition metal silicate crystals grow upwards as shown.

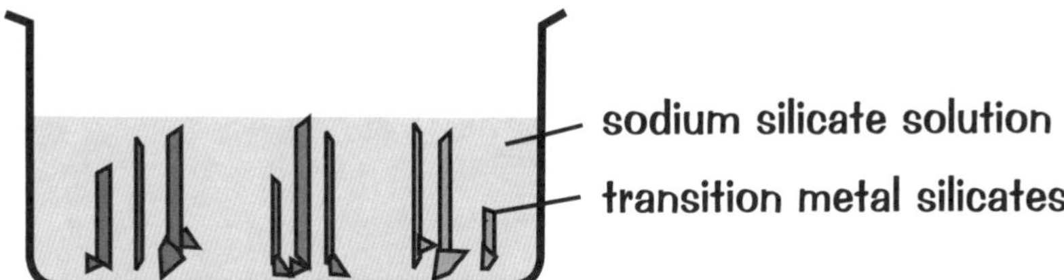

a) Circle the three colours that you would be likely to see in the garden if potassium chromate(VI), potassium manganate(VII) and copper(II) sulfate crystals are used.

red orange yellow green blue purple

b) Cobalt(II) chloride produces pink cobalt silicate crystals.

i) What is the electron configuration of a cobalt atom? ....

ii) Which cobalt ion is present in cobalt(II) chloride? ....

Q5 Read the description of **metal X** and answer the question that follows.

'Metal X is found in the block of elements between groups II and III in the periodic table. It has a melting point of 1860 °C and a density of 7.2 g/cm^3. The metal is used to provide the attractive shiny coating on motorbikes and bathroom taps. The metal forms two coloured chlorides, XCl_2 (blue) and XCl_3 (green).'

Identify six pieces of evidence in the passage which suggest that metal X is a transition metal.

1. ....
2. ....
3. ....
4. ....
5. ....
6. ....

Top Tips: Most of the first 10 transition metals have two electrons in the 4th energy level. Chromium and copper are a little bit different — they only have one electron in the 4th energy level.

Industrial Salt

Q1 Indicate whether the following statements about obtaining salt are **true** or **false**.

		True	False
a)	In the UK most salt is obtained by evaporation in flat open tanks.	☐	☐
b)	There are massive deposits of rock salt in the Lake District and Kent.	☐	☐
c)	Salt can be mined by pumping hot water underground.	☐	☐
d)	Rock salt is a mixture of salt and impurities.	☐	☐

Q2 Describe three ways that **rock salt** is used.

1. ....

2. ....

3. ....

Q3 **Circle** the correct answer for each of the questions below.

a) One of the products of the electrolysis of brine is chlorine. You can test for it by:

Using a glowing splint — chlorine will relight it.

Using damp litmus paper — chlorine will bleach it.

Using universal indicator — chlorine will turn it purple.

b) Two of the products of electrolysis are reacted together to make household bleach. They are:

Chlorine and hydrogen.

Chlorine and sodium hydroxide.

Hydrogen and sodium hydroxide.

Q4 Three very useful products, **chlorine**, **hydrogen** and **sodium hydroxide**, are produced by the electrolysis of brine.

Tick the boxes to show which of these three products is used for each purpose listed below.

		Chlorine	Hydrogen	Sodium hydroxide
a)	Used in oven cleaner	☐	☐	☐
b)	Used to kill bacteria	☐	☐	☐
c)	Used in the Haber process	☐	☐	☐
d)	Used to make PVC	☐	☐	☐
e)	Used to make fats from oils	☐	☐	☐

Industrial Salt

Q5 Harry runs a little **brine electrolysis** business from his garden shed. He keeps a record of all the different **industries** that he sells his products to.

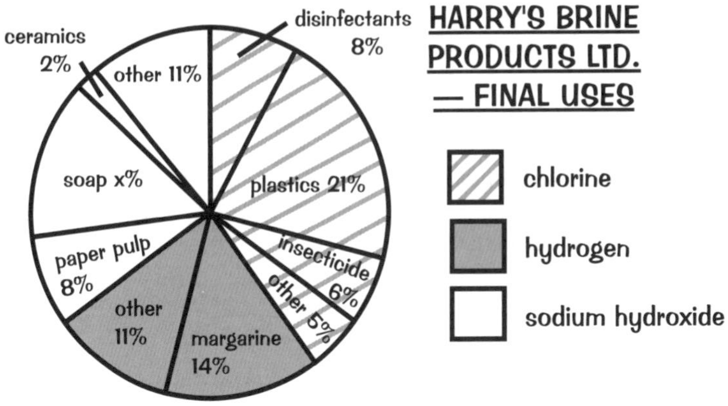

HARRY'S BRINE PRODUCTS LTD. — FINAL USES

a) Which brine product, **hydrogen**, **chlorine** or **sodium hydroxide**, does Harry sell the most of?

....

b) What percentage of Harry's products are used to manufacture **soap**?

....

c) Which **industry** uses the biggest proportion of Harry's products? ....

Q6 When either **dilute** brine or **molten** sodium chloride are electrolysed, the products are different to those made from **concentrated** brine.

a) Complete the table to show what products are formed at each electrode.

	Product formed at:	
	Anode	Cathode
concentrated brine	chlorine	hydrogen
dilute brine		
molten sodium chloride		

b) Write the equations for the reactions happening at the electrodes with **dilute brine**.

Anode: .... Cathode: ....

c) Why are the reactions at the electrodes different with **dilute** brine than with concentrated brine?

....

....

d) Write the equations for the reactions happening at the electrodes with **molten sodium chloride**.

Anode: .... Cathode: ....

Top Tips: Well, I bet you really didn't want to know that much about salt. In fact, I reckon you would have been happy just knowing that you can put it on your chips. But unfortunately, you're going to have to know all of its industrial uses AND what products you get when you electrolyse concentrated brine, dilute brine and molten sodium chloride. It's tough, but someone's got to do it.

CFCs and the Ozone Layer

Q1 **CFCs** are a **useful** group of chemicals.

a) Which of the following shows the structure of a CFC? Circle your answer.

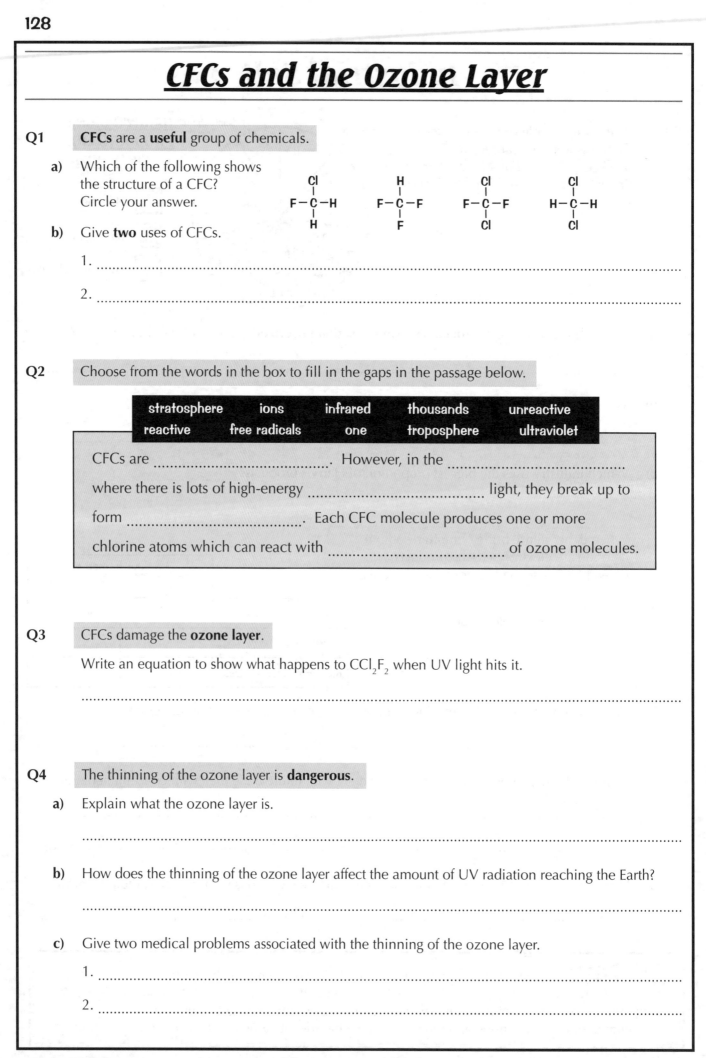

Cl (top), F–C–H, H (bottom)	H (top), F–C–F, F (bottom)	Cl (top), F–C–F, Cl (bottom)	Cl (top), H–C–H, Cl (bottom)

b) Give **two** uses of CFCs.

1. ....

2. ....

Q2 Choose from the words in the box to fill in the gaps in the passage below.

stratosphere	ions	infrared	thousands	unreactive
reactive	free radicals	one	troposphere	ultraviolet

CFCs are .... However, in the .... where there is lots of high-energy .... light, they break up to form .... Each CFC molecule produces one or more chlorine atoms which can react with .... of ozone molecules.

Q3 CFCs damage the **ozone layer**.

Write an equation to show what happens to CCl_2F_2 when UV light hits it.

....

Q4 The thinning of the ozone layer is **dangerous**.

a) Explain what the ozone layer is.

....

b) How does the thinning of the ozone layer affect the amount of UV radiation reaching the Earth?

....

c) Give two medical problems associated with the thinning of the ozone layer.

1. ....

2. ....

CFCs and the Ozone Layer

Q5 **Free radicals** are very reactive particles.

a) Draw a dot and cross diagram in the box to show the covalent bond between the two hydrogen atoms in a hydrogen molecule.

b) What type of particle is formed when a covalent bond breaks:

i) unevenly (i.e. both electrons go to the same atom)? ....

ii) evenly (i.e. one electron goes to each atom)? ....

c) Give the symbol for a hydrogen free radical. ....

d) What makes free radicals so reactive? ....

Q6 **Free radicals** found in the upper atmosphere are responsible for the depletion of the ozone layer.

a) Write an equation to show what happens when a chlorine free radical hits an ozone molecule.

....

b) The chlorine oxide (ClO•) free radical produced will then react with another ozone molecule. Write an equation to show this reaction.

....

c) How many ozone molecules are destroyed in these two reactions? ....

Q7 It's now clear that it was a mistake to use CFCs without fully understanding the effects they could have. Many countries have now **banned** the use of CFCs.

a) Explain why the effects of CFCs are a global problem.

....

b) Explain why even a complete ban will not stop the damage to the ozone layer.

....

....

Q8 **Replacements** for CFCs are being developed.

a) Circle two substances from the list below that are thought to be suitable replacements for CFCs.

chlorocarbons | alkanes | trichlorides | alkenes | hydrofluorocarbons

b) One of the substances in the above list is a group of compounds that are very similar to CFCs. What is the important difference that makes these compounds safe to use?

....

Answers

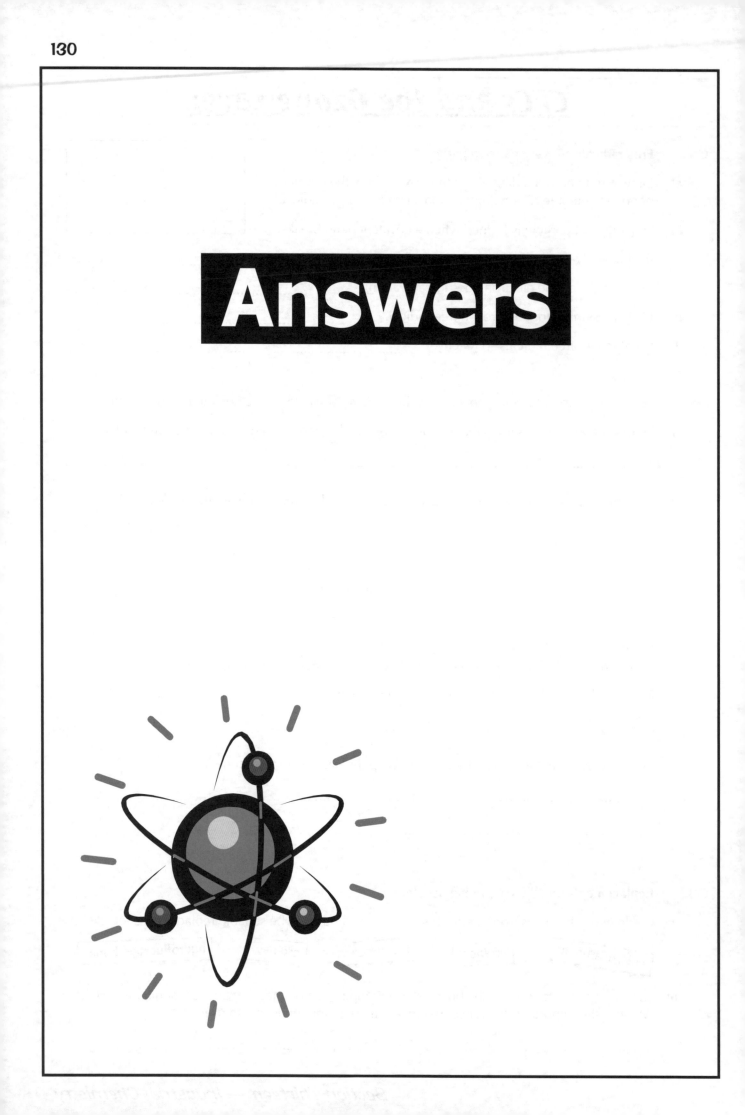

Section One — Chemical Concepts

Section One — Chemical Concepts

Page 1 — Atoms

Q1

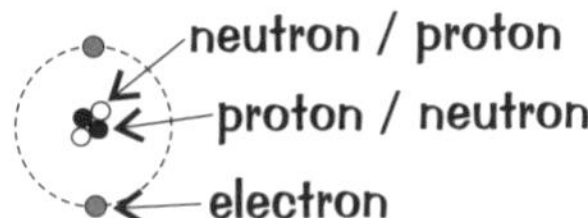

Q2a) protons
b) zero / 0
c) protons/electrons, electrons/protons

Q3

Particle	Mass	Charge
Proton	1	**+1**
Neutron	1	0
Electron	**1/2000**	–1

Q4a) 24 — the mass number tells you the total number of protons and neutrons in the nucleus.
b) 12 — the atomic number tells you the number of protons in the nucleus.
c) 24 – 12 = 12 neutrons

Page 2 — Solids, Liquids and Gases

Q1a) gas
b) solid
c) liquid
d) gas
e) gas
f) liquid

Q2 moving, attraction, speeds, quickly, evaporation

Q3a) It increases because the particles gain energy and vibrate more. This causes them to move apart slightly, increasing the volume that they take up.
b) It increases because the particles gain energy and move faster. They exert a greater pressure because they bounce off the walls of the container more often and with more force.

Q4 As the temperature increases, more liquid particles gain enough energy to overcome the forces of attraction keeping them together and become gas particles. These can then move about the room and be detected by the nose.

Page 3 — Elements, Compounds and Mixtures

Q1a) True
b) False
c) False
d) True

Q2 compounds, separate, different, separate, bonds

Q3 copper, oxygen

Q4 E.g. all the atoms in gold and aluminium are the same, but in water and sugar there are atoms of more than one element.

Page 4 — The Periodic Table

Q1a) vertical
b) metals
c) transition
d) right-hand
e) similar

Q2 Any two of: neon, krypton, helium, xenon, radon.

Q3a) increases
b) decreases

Q4a) True
b) False
c) True
d) False

Pages 5-6 — Balancing Equations

Q1a) Correctly balanced
b) Incorrectly balanced
c) Incorrectly balanced
d) Correctly balanced
e) Correctly balanced

Q2 $2C + O_2 \rightarrow 2CO$

Q3a) Reactants: methane and oxygen, Products: carbon dioxide and water.
b) methane + oxygen → carbon dioxide + water
c) $CH_4 + 2O_2 \rightarrow CO_2 + 2H_2O$

Q4a) $2Na + Cl_2 \rightarrow 2NaCl$
b) $4Li + O_2 \rightarrow 2Li_2O$
c) $MgCO_3 + 2HCl \rightarrow MgCl_2 + H_2O + CO_2$
d) $2Li + 2H_2O \rightarrow 2LiOH + H_2$

Q5a) $CuO + 2HBr \rightarrow CuBr_2 + H_2O$
b) $H_2 + Br_2 \rightarrow \mathbf{2}HBr$
c) $\mathbf{2}Mg + O_2 \rightarrow 2MgO$
d) $2NaOH + H_2SO_4 \rightarrow Na_2SO_4 + \mathbf{2}H_2O$

Q6a) $\mathbf{3}NaOH + AlBr_3 \rightarrow \mathbf{3}NaBr + Al(OH)_3$
b) $\mathbf{2}FeCl_2 + Cl_2 \rightarrow \mathbf{2}FeCl_3$
c) $N_2 + \mathbf{3}H_2 \rightarrow \mathbf{2}NH_3$
d) $\mathbf{4}Fe + \mathbf{3}O_2 \rightarrow \mathbf{2}Fe_2O_3$
e) $\mathbf{4}NH_3 + \mathbf{5}O_2 \rightarrow \mathbf{4}NO + \mathbf{6}H_2O$

Section Two — Products from Rocks

Page 7 — Using Limestone

Q1a) calcium carbonate
b) i) calcium oxide
ii) $CaO + H_2O \rightarrow Ca(OH)_2$
iii) Adding to fields to neutralise acidic soils (it is alkaline).

Q2 granite, paint

Section Two — Products from Rocks

Q3a) 1. calcium oxide (quicklime) — CaO
2. carbon dioxide — CO_2

b) magnesium oxide

c) $CuCO_3 \rightarrow CuO + CO_2$

Q4 Any three of, e.g. quarrying uses up land / spoils natural landscapes / causes noise and dust / increased traffic (which causes noise and air pollution) / the waste produces unsightly tips.

Page 8 — Metals from Rocks

Q1a) A mineral which contains enough metal to make extracting the metal from it worthwhile.

b) oxygen and sulfur

Q2 Advantage, e.g. useful products can be made / it provides local people with jobs / it brings money to an area.
Disadvantage, e.g. it causes noise / scarring of the landscape / destruction of habitats.

Q3a) It is too impure to conduct electricity well.

b) Impure copper (obtained e.g. by reduction with carbon) is purified using electrolysis.

c) e.g. water pipes, making coins

Q4 Any two from, e.g. there is a limited supply of copper. There is an increasing demand for copper. To reduce the amount of material going into landfill. Because it requires less energy than extracting new copper.

Page 9 — Extraction of Metals

Q1a) Because zinc is more reactive than copper.

b) i) No

ii) Zinc is less reactive than aluminium so it wouldn't be able to push the aluminium out and bond to the oxygen.

Q2a) Carbon (in the wood) is more reactive than copper, so it 'steals' oxygen from the copper ore.

b) Copper is very unreactive so the reaction would happen at a relatively low temperature.

Q3 carbon, below, reduction, electrolysis, more.

Q4 The order of reactivity, from most to least, is: dekium, candium, bodium, antium.

Pages 10-11 — Properties of Metals

Q1a) the transition metals

b) They conduct electricity well (and are not very reactive).

Q2a) i) N — It is stronger and lighter.

ii) M — It is more resistant to corrosion.

iii) N — It has the higher melting point.

b) Its very high density.

Q3a) malleable

b) conducts heat

c) resists corrosion

d) ductile

Q4 Desirable qualities in a metal used to make knives and forks would be: strong, resistant to corrosion, visually attractive/shiny, non-toxic.

Q5a) A and B

b) A, because it took the least time for the end that wasn't near the heat source to heat up.

Q6a)

free electrons
metal atoms

b) There are a lot of free electrons in a metal (because of the way the atoms are bonded together).

Q7 Any three from: good conductor of electricity, good conductor of heat, strong, easily bendable, malleable.

Pages 12-13 — More Metals

Q1a) copper

b) nickel

c) silver, copper or nickel

Q2a) False

b) True

c) False

d) False

e) True

f) False

g) True

Q3a) A mixture of two or more metals or a mixture of a metal and a non-metal.

b) By adding small amounts of carbon or other metals to the iron.

Q4

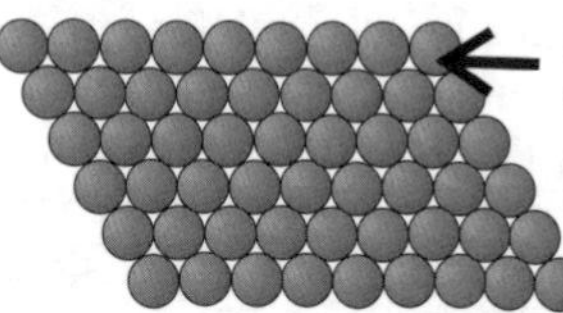

layers of atoms can slide over each other so the iron is easy to bend and shape

Q5a) 37.5% ($9 \div 24 \times 100 = 37.5$)

b) 9-carat gold is harder than pure gold because it is an alloy and so it contains different sized atoms. The atoms in 9-carat gold can't slide over each other as easily as the ones in pure gold can.

Q6a) E.g. in aeroplane wings, for making spectacle frames.

b) They can return to their original shape if they have been bent out of shape.

c) They are currently very expensive. They suffer more from metal fatigue than ordinary metals.

Section Three — Products from Crude Oil

Page 14 — Paints and Pigments

Q1 pigment — gives paint its colour
colloid — tiny particles dispersed in another material
solvent — keeps paint runny
binding medium — holds pigment particles to a surface

Q2a) oil-based, water-based
b) solvent
c) something that dissolves oil

Q3a) i) oil-based
ii) Some of the solvents may produce harmful fumes. You should make sure there is plenty of ventilation.
b) Water-based. Most of the paint sold is used for internal decoration, and you would be more likely to use a water-based paint for this.

Q4a) False
b) False
c) True
d) True

Section Three — Products from Crude Oil

Page 15 — Fractional Distillation of Crude Oil

Q1a) mixture
b) aren't
c) last
d) larger

Q2

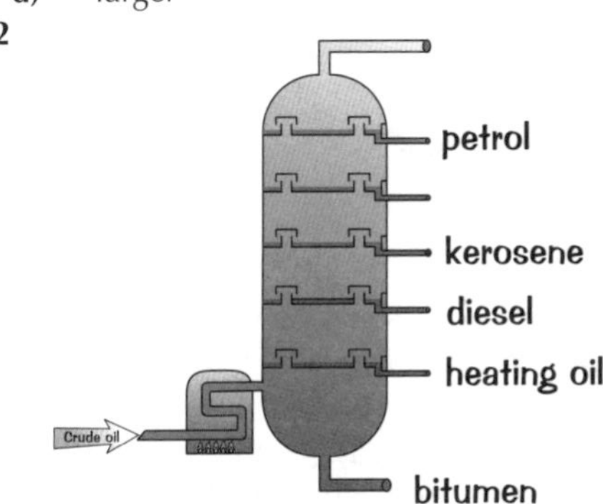

Q3 The larger the molecule, the higher the boiling / condensing point.

Q4a) Petrol has a lower boiling point.
b) Petrol is more flammable / sets on fire more easily.
c) Petrol is less viscous / flows more easily.
d) Petrol is more volatile / evaporates more readily.

Page 16 — Burning Hydrocarbons

Q1a) hydrogen, carbon
b) Fuels are substances that react with **oxygen** to release **energy**.
c) Incomplete combustion happens if there is not enough oxygen available.

Q2a) blue
b) hydrocarbon + oxygen → carbon dioxide + water
c) i) $CH_4 + \mathbf{2O_2} \rightarrow \mathbf{CO_2} + \mathbf{2H_2O}$
ii) $C_3H_8 + \mathbf{5O_2} \rightarrow \mathbf{3CO_2} + \mathbf{4H_2O}$
d) The water pump draws the gases produced by the burning hexane through the tube. The water vapour cools and turns back into liquid in the section with the ice, and you can show it's water by checking its boiling point. The limewater turns milky, showing that CO_2 was also made.

Q3a) E.g.
$C_4H_{10} + \mathbf{4O_2} \rightarrow \mathbf{5}H_2O + CO_2 + \mathbf{CO} + \mathbf{2C}$
b) i) Because the carbon monoxide produced is poisonous.
ii) Because the fuel releases less energy if it burns incompletely.
iii) Because carbon is produced in the form of soot.
c) During incomplete combustion the flame is yellow not blue, and smoky (from the soot).

Page 17 — Using Crude Oil as a Fuel

Q1 E.g. petrol, diesel, kerosene, and gases such as propane and butane.

Q2a) Solar power doesn't work in dark places / at night.
b) The wind wouldn't always be blowing when someone wanted to use the oven.
c) Nuclear fuels (and radioactive waste) would be very dangerous if they leaked. / It would be very expensive to develop a way of storing the fuel / disposing of waste in a safe way.

Q3a) There is the possibility of spills into the sea. Crude oil is harmful to seabirds and to many other sea creatures.
b) Burning oil products releases substances that cause acid rain, global warming and global dimming.

Q4 E.g. New reserves of oil have been discovered since the 1960s. New methods of extraction mean that oil that was once too expensive or difficult to extract can now be used.

Q5 E.g. Most vehicles, heating systems, etc. around today are designed to use crude oil fractions as fuel, and converting to alternatives would be time-consuming and costly. We use more energy than can currently be supplied by the available alternatives to crude oil products.

Section Four — Food and Carbon Chemistry

Page 18 — Alkanes and Alkenes

Q1

```
   H        H H          H H H          H H H H
   |        | |          | | |          | | | |
 H-C-H    H-C-C-H      H-C-C-C-H      H-C-C-C-C-H
   |        | |          | | |          | | | |
   H        H H          H H H          H H H H
methane    ethane      propane         butane
```

Q2a) C_nH_{2n+2}
b) $C_{20}H_{42}$ (n = 20 so 2n + 2 = (2 × 20) + 2 = 42)
Q3a) C_2H_4
b)

```
H     H
 \   /
  C=C
 /   \
H     H
```

c) Propene
d)

```
     H  H
H    |  |
 \   C=C-C-H
  C=C  ...
```

```
H     H  H
 \    |  |
  C = C- C - H
 /       |
H        H
```

Q4a) C_5H_{10}
b) C_6H_{12}
c) C_8H_{16}
d) $C_{12}H_{24}$
Q5a) False
b) True
c) False
d) False
e) True

Page 19 — Cracking Crude Oil

Q1 shorter, petrol, diesel, long, high, catalyst, molecules, cracking
Q2a) E.g. petrol, paraffin, ethene
b) They are too thick (viscous) / They don't burn as easily as shorter hydrocarbons.
c) thermal decomposition
Q3
1. The long-chain molecules are heated.
2. They are vaporised (turned into a gas).
3. The vapour is passed over a catalyst at a high temperature.
4. The molecules are cracked on the surface of the catalyst.

Q4a) decane → octane + ethene
b) $C_{10}H_{22} \rightarrow C_8H_{18} + C_2H_4$

Page 20 — Making Polymers

Q1 The monomer of polyethene is ethene.
Q2a) They contain at least one double covalent bond.
b) High pressure and a catalyst.
Q3a)

$$n\begin{pmatrix} CH_3 & H \\ | & | \\ C & = C \\ | & | \\ H & H \end{pmatrix} \rightarrow \begin{pmatrix} CH_3 & H \\ | & | \\ -C & - C- \\ | & | \\ H & H \end{pmatrix}_n$$

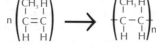

b) polypropene
Q4a) Ruler 2
b) The forces between the molecules are weaker in ruler 1, which allows the molecules to slide over one another and separate more easily.

Page 21 — Uses of Polymers

Q1 waterproof, lightweight
Q2a) Plastics don't decay, so the landfill sites soon fill up. / This is a waste of land and of plastic.
b) Some plastics give off poisonous gases like hydrogen chloride and hydrogen cyanide when they are burned.
c) The different types of plastic have to be separated out before they can be recycled. This is difficult and expensive.
Q3a) The coating of polyurethane makes the jacket waterproof.
b) The jacket made from breathable fabric, because it is waterproof but it also allows water vapour / sweat to escape, which is more comfortable during exercise.
Q4a) kettles
b) disposable cups
c) carrier bags
d) window frames

Section Four — Food and Carbon Chemistry

Page 22 — Chemicals and Food

Q1a) Heat energy breaks some of the chemical bonds in the protein molecule. The protein molecules then change shape.
b) Denaturing the proteins changes the texture of the food and makes it more appealing — less slimy (eggs) or chewy (meat).
Q2 carbohydrate, cellulose, digest, heat
Q3a)

hydrophobic

hydrophilic

b) Lecithin molecules surround the droplets of oil, with their hydrophilic heads facing out into the water and their hydrophobic tails in the oil droplet. This layer keeps the oil droplets from joining together to separate out from the water.

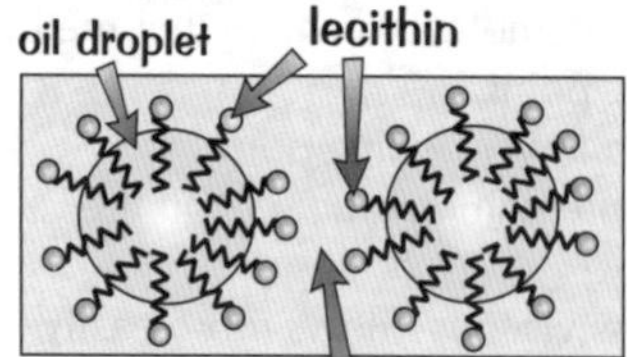

Section Four — Food and Carbon Chemistry

Page 23 — Packaging and Smart Materials

Q1a) E.g. on drinks cans and bottles to show when they're cold enough to drink / at the best temperature to drink / at the optimum temperature to drink.
b) E.g. in car airbag sensors.
c) E.g. in sunglasses.
Q2a) Foods often 'go off' because of mould or bacteria growing on them. Most bacteria and moulds can't grow without water.
b) A sachet of a desiccant such as silica gel can be added to the packet to absorb water.
Q3a) The coloured dot contains a dye which changes colour faster the warmer it gets. It shows if a food has been warm enough for long enough to allow microbes to grow.
b) yes
c) E.g. any two of: How quickly the chicken goes off will depend on what temperature it is kept at / how fresh it was when it was packaged / how fresh it was when it was bought.

Page 24 — Food Additives

Q1a) oxygen
b) antioxidants / preservatives
Q2a) E.g. cochineal or gelatin
b) Because they are allergic to it.
Q3a) chromatography
b) A
Q4a) The control group would not take additive X (though they might be given a placebo), but would still be monitored to see how often they suffered from migraines. This group is needed to give an idea of the percentage of people suffering migraines normally, to see if X makes a difference.
b) Increasing the number of people monitored in the research.
c) There's an increase in the incidence of migraines in the male group taking substance X — 11% is significantly greater than the control result of 3%. There's no significant difference between the two female groups.
d) There may be a link, but it doesn't prove the additive caused the increased incidence of migraines in the male group.

Page 25 — Plant Oils in Food

Q1a) Fruits: e.g. avocados and olives
Seeds: e.g. brazil nuts and sesame seeds
b) e.g. food or fuel
c) It squeezes the oil out of the plant material.
Q2 Saturated animal fat — bottom diagram
Polyunsaturated grape seed oil — top diagram
Monounsaturated olive oil — middle diagram
Q3a) Reaction with hydrogen with a nickel catalyst at about 60 °C. The double bonds open up and bond to the hydrogen atoms.
b) It increases the melting points of vegetable oils.
Q4 unsaturated, saturated, less, increase, cholesterol

Pages 26-27 — Plant Oils as Fuel

Q1a) E.g. rapeseed oil, soy bean oil
b) They contain a lot of energy.
Q2 $0.9 \times 37\ 000\ 000 =$ **33 300 000 J** or 33.3 MJ
Q3a) Any three of:
Burning biodiesel produces no net increase in carbon dioxide.
It produces less sulfur dioxide pollution / particulates.
It is biodegradable and less toxic.
It conserves crude oil, which is a non-renewable resource.
b) We can't make enough of it to replace all the petrol and diesel used in the UK / It costs more to make and so it tends to be more expensive.
c) Any two from:
Biodiesel can be used in diesel car engines without the engine having to be modified.
Biodiesel can be supplied using existing filling stations and pumps.
Biodiesel is less flammable.
Biodiesel doesn't have to be kept under pressure. (Other answers are possible.)
Q4a) Biodiesel does produce carbon dioxide when burnt, but because it comes from recently grown plants which took in this carbon dioxide during their lives it does not increase the net level of the gas in the atmosphere.
b) Normal diesel is made from crude oil — the remains of dead plants and animals from millions of years ago. When it is burnt it produces a net increase in the level of carbon dioxide.
Q5a) Recycled cooking oil
b) Climate change might stop increasing so fast. Spills would be less harmful to the environment.
c) It has reduced the tax on biodiesel and increased the tax on normal diesel.
d) The Government would get less money from fuel tax. It might have to make cuts in other places (e.g. education) or raise the tax on something else.
e) You don't need to get a diesel car modified. Biodiesel may cost slightly more but the Government is actually making less money on it than it does on normal diesel.

Section Five — The Earth and the Atmosphere

Page 28 — Ethanol

Q1 Temperature of 300 °C and a pressure of 70 atmospheres.

Q2 sugars, enzymes, temperature, low, high, enzymes, 30 °C, oxygen, concentration, 10–20%

Q3a) A

b) C

Q4a) Brazil — they have suitable land and a suitable climate for growing plant material that can be fermented to give ethanol. (Other countries could be suggested)

b) It reduces the amount of fossil fuel that is burnt and the crops used to make it absorb some carbon dioxide when they are growing.

Page 29 — Perfumes

Q1a) Animal testing helps to make sure that a chemical isn't poisonous and won't burn or irritate the skin before it is used on humans.

b) Any one of:
Testing may cause the animals pain and suffering.
The animals have no choice about the testing.
Chemicals may not affect humans in the same way as the test animal, making the test useless.

Q2a) Compound C, because it won't react with sweat or wash off easily, and it evaporates at a low enough temperature.

b) The chemical must be tested to make sure that it is non-toxic and that it does not irritate the skin.

Q3a)
1. Put 15 cm^3 of ethanoic acid into a 100 cm^3 conical flask.
2. Add 15 cm^3 of ethanol and a few drops of concentrated sulfuric acid.
3. Warm the flask gently on an electric heating plate for 10 minutes.
4. Turn off the heat.
5. When the flask is cool enough to handle, pour its contents into a 250 cm^3 beaker containing 100 cm^3 of sodium carbonate solution.

b) To neutralise the solution.

c) It is a catalyst.

Section Five — The Earth and the Atmosphere

Page 30 — The Earth's Structure

Q1 Diagram should show a sphere with 3 layers.
Labels:
Crust (outer layer) — very thin, has an average thickness of 20 km.
Mantle (next layer down) — properties of a solid but flows very slowly like a liquid.
Core (centre) — mostly iron and nickel.

Q2 The main earthquake zones are along the plate boundaries.

Q3

	Evidence	How reliable is it?
Earthquake	Strain in underground rocks.	Low reliability, can only suggest the possibility of an earthquake
Volcanic Eruption	Rising molten rock causing the ground to bulge slightly, leading to mini earthquakes.	Low/medium reliability, molten rock can cool instead of erupting, not a definite sign

Page 31 — Evidence for Plate Tectonics

Q1 Pangaea, continents, plate tectonics, living creatures, fossils (last two can be in either order)

Q2a) The continents are too far apart for seeds to spread this way.

b) That the continents were once joined and have since separated.

Q3a) A and C — they have identical rock sequences.

b) Similar animal and plant fossils in the rocks.

c) E.g. The coastlines of South America and South Africa seem to fit together like a jigsaw.
Living creatures, such as a particular earthworm, are found on both sides of the Atlantic Ocean.

Pages 32 — The Three Different Types of Rock

Q1 igneous rocks — formed when magma cools — granite
metamorphic rocks — formed under intense heat and pressure — marble
sedimentary rocks — formed from layers of sediment — limestone

Q2 Igneous, crust, slowly, big, intrusive, granite, gabbro, quickly, small, extrusive, basalt, rhyolite

Q3a) E.g. the church is made from limestone which is formed mostly from sea shells.

b) The pressure forces out the water. Fluids flowing through the pores deposit minerals that cement the sediment together.

c) They are both made from the same chemical — calcium carbonate.

Section Five — The Earth and the Atmosphere

Pages 33-34 — The Evolution of the Atmosphere

Q1 True statements:
When the Earth was formed, its surface was molten. When some plants died and were buried under layers of sediment, the carbon they had removed from the atmosphere was locked up as fossil fuels.

Q2 The statements should be in this order (from the top of the timeline):
1. The atmosphere is about four-fifths nitrogen and one-fifth oxygen.
2. More complex organisms evolved.
3. Oxygen built up due to photosynthesis and the ozone layer developed.
4. Plant life appeared.
5. Water vapour condensed to form oceans.
6. The Earth cooled down slightly. A thin crust formed.
7. The Earth formed. There was lots of volcanic activity.

Q3 The amount of carbon dioxide has massively decreased because it dissolved into the oceans and green plants used it up in photosynthesis. Some then became locked up in fossil fuels.

Q4a) Largest sector (pale grey) is nitrogen, second largest (dark grey) is oxygen, smallest (blue) is carbon dioxide and other gases.

b) Nitrogen: 80% approx. (it's 78% in dry air)
Oxygen: 20% approx. (it's 21% in dry air)

c) Nitrogen has increased. Carbon dioxide has decreased. There's far less water vapour now. Oxygen now makes up a significant proportion of the atmosphere.

d) As the planet cooled, the water vapour condensed and formed the oceans.

e) Plants and microorganisms photosynthesised and produced it.

f) In any order:
Created the ozone layer which blocked harmful rays from the Sun.
Killed off early organisms / allowed more complex organisms to evolve.

Page 35 — Atmospheric Change

Q1 The water in comets has a higher proportion of 'heavy' water than the water found on Earth.

Q2a) burning fossil fuels, deforestation

b) E.g. as CO_2 levels rise, the trend in global temperature is also upwards.

Q3a) The ozone layer protects us from harmful UV radiation which causes skin cancer. As the ozone layer has thinned, this has increased our exposure to UV.

b) No. The rise could also be due to, for example, more people sunbathing or taking holidays in sunny countries.

Page 36 — The Carbon Cycle and Climate Change

Q1 True statements:
The greenhouse effect is needed for life on Earth as we know it.
Increasing amounts of greenhouse gases lead to global warming.

Q2a) burning

b) photosynthesis

c) respiration / burning

d) coal

Q3a) The greenhouses gases absorb much of the heat radiated away from Earth and re-radiate it in all directions, including back towards the Earth. This keeps the atmosphere relatively warm.

b) More greenhouse gas in the atmosphere means that more of the Sun's heat is trapped rather than radiated back out into space. This means that the Earth gets warmer.

Q4 A: decay (also accept respiration)
B: photosynthesis
C: animal eats plant / feeding
D: you eat animal

Page 37 — Human Impact on the Environment

Q1a) The human population is now **bigger** than it was 1000 years ago.

b) The growth of the human population is now **faster** than it was 1000 years ago.

c) The human impact on the environment is now **greater** than it was 1000 years ago.

Q2a) i) Water: E.g. 2 from discharging sewage, fertilisers washing in from agricultural land, oil spills from tankers, nuclear waste, toxic chemicals from industries, gases from power stations cause acid rain that pollutes water.

ii) Land: E.g. 2 from fertilisers used in agriculture, household waste, nuclear waste, toxic chemicals from industry and agriculture.

iii) Air: E.g. 2 from gases released from power stations and industries, smoke, sewage/ household waste gives off unpleasant smells.

b) E.g. 2 from: building, farming, quarrying.

Section Five — The Earth and the Atmosphere

Q3a) i) John is more likely to live in the UK city and Derek in rural Kenya.

ii) The following should be ticked:
John buys more belongings, which use more raw materials to manufacture.
John has central heating in his home and Derek has a wood fire.
John drives a car and Derek rides a bicycle.

b) Any sensible suggestion, such as: John could use his car less, use his central heating less, recycle more waste, buy fewer new things, etc.

Page 38 — Air Pollution and Acid Rain

Q1 sulfur dioxide, sulfuric, nitrogen oxides, nitric

Q2a) If acid rain falls, limestone will react with the acid and wear away.

b) E.g. It damages plants / corrodes metal / acidifies lakes and kills fish, etc.

c) Flue Gas Desulfurisation technology can be installed — this removes sulfur dioxide from the gases released.

Q3a) In an engine it's more likely that there wouldn't be enough oxygen, causing incomplete combustion.

b) It prevents blood carrying oxygen around the body, which can be fatal.

Q4a) carbon monoxide + nitrogen oxide → **nitrogen + carbon dioxide**

b) $2CO + 2NO \rightarrow N_2 + 2CO_2$

c) A mixture of platinum and rhodium.

Pages 39-40 — Protecting the Atmosphere

Q1a) carbon dioxide and water

b) The production of biogas is slow at low temperatures.

Q2a)

Fuel	Initial Mass (g)	Final Mass (g)	Mass of Fuel Burnt (g)
A	98	92	6
B	102	89	13

b) fuel A

c) E.g. any four of: availability, ease of storage, cost, toxicity, ease of use, amount of pollution caused

Q3a) Thousands of monitoring stations all over the world collect data about conditions on Earth. Computers are then used to carry out millions of calculations and generate predictions from the data.

b) Computer models have to use assumptions and will only give accurate predictions if these are correct. Not enough is known about some aspects of climate to guarantee this. The input data is very incomplete. Also, one small error in the data or calculations could have a big impact on the model.

c) By comparing the predictions of the model with what actually happens as time goes on.

Q4a) The 'precautionary principle' says that whoever suggests an action needs to show that it isn't harmful. But the harm it might do needs to be weighed up against the harm that'll be caused by doing nothing.

b) Strategies to reduce climate change have disadvantages, but these need to be weighed up against the harm that'll be caused by doing nothing and allowing climate change to continue.

c) Any two of:
Burn less fossil fuel OR examples of how to do this, e.g. burn fossil fuels more efficiently, use other sources of energy that don't produce greenhouse gases such as solar or wind power.
Capture CO_2 produced by burning fossil fuels before it's released into the atmosphere.
Plant forests to absorb some of the CO_2.

Q5a) water

b) When hydrogen is burned no carbon dioxide is produced so it doesn't contribute to global warming. It doesn't release particulates either, and it doesn't produce sulfur dioxide so doesn't cause acid rain.

c) E.g. Hydrogen-powered vehicles need very expensive engines. Hydrogen is difficult to store safely. Fuelling stations would need to be adapted /converted. Producing hydrogen needs a lot of energy, and generating this can be polluting.

Page 41 — Recycling Materials

Q1 Sustainable development means meeting the needs of today's population without harming the ability of future generations to meet their own needs.

Q2 Any three of: less finite resources are used up. / Recycling some materials (e.g. glass) uses a fraction of the energy that mining, extracting and purifying the original material would. / Using less energy means that less pollution is caused by burning fossil fuels to generate it. / Using less energy saves money. / Less rubbish has to be put into landfill, which takes up space and pollutes the surroundings.

Q3 E.g. paper that's just thrown away will take up space in landfill sites. / It takes more energy to produce new paper.

Q4a) i) $4 \times 1 = 4$ tonnes

ii) 3 billion × 20 = 60 billion g of aluminium
60 billion ÷ 1000 = 60 million kg of aluminium
60 million ÷ 1000 = 60 000 tonnes of aluminium

iii) 60 000 × 4 = 240 000 tonnes of bauxite

b) i) Deforestation may be required to set up the mines as bauxite is often found in rainforests. The mining itself can cause pollution and ruin the landscape.

ii) Large amounts of electricity are needed to extract the aluminium. Producing electricity using fossil fuels means adding to atmospheric CO_2 levels.
iii) This means more bauxite has to be mined and the aluminium extracted, destroying more rainforest and producing more CO_2. The waste cans would also increase the amount of landfill.

Section Six — Classifying Materials

Page 42 — The Periodic Table and Electron Shells

Q1a) radon and krypton
b) silicon and sodium
c) sodium
d) nickel
e) iodine
f) silicon or iodine
Q2a) True
b) False
c) True
d) False
e) True
Q3a) The following should be ticked: **A** and **D**
b) Fluorine and chlorine are both in Group VII, so they both contain the same number of electrons in their outer shell. The chemistry of elements is decided by the number of electrons they have in their outer shell.
Q4

	Alternative Name for Group	Number of Electrons in Outer Shell
Group 1	**Alkali metals**	**1**
Group 7	**Halogens**	**7**
Group 8	**Noble gases**	**8**

Pages 43-44 — Electron Shells

Q1a) i) True
ii) False
iii) False
iv) True
v) False
b) ii) The lowest energy levels are always filled first.
iii) Atoms are most stable when they have full outer shells.
v) Reactive elements have partially filled outer shells.
Q2 E.g. The innermost electron shell should be filled first / there should be two electrons in the inner shell.
The outer shell contains too many electrons / should only hold a maximum of eight electrons.

Q3a) 2,2
b) 2,6
c) 2,8,4
d) 2,8,8,2
e) 2,8,3
f) 2,8,8
Q4a) Noble gases are unreactive elements because they have full outer shells of electrons.
b) Alkali metals are reactive elements, because they have incomplete outer shells of electrons.
Q5a) 2,8,7
b)

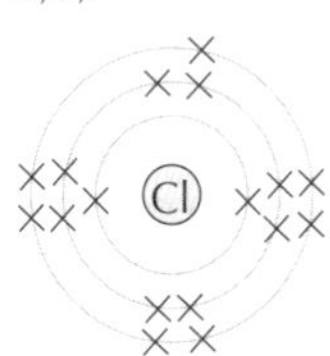

c) It only needs one electron in order to have a full outer shell, so is keen to react and get one.
Q6

a) C b) N c) F
d) Na e) Mg f) P
g) S h) K i) Ca

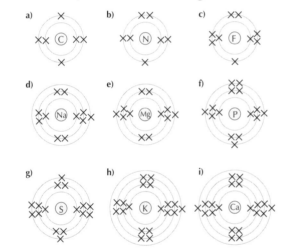

Page 45-46 — Ionic Bonding

Q1a) electrons, ions
b) charged particles
c) attracted to
Q2a) True
b) False
c) True
d) True
e) False
Q3a) Group I
b) one
c) +1
d) NaCl
Q4 BeS, K_2S, BeI_2, KI
Q5a)

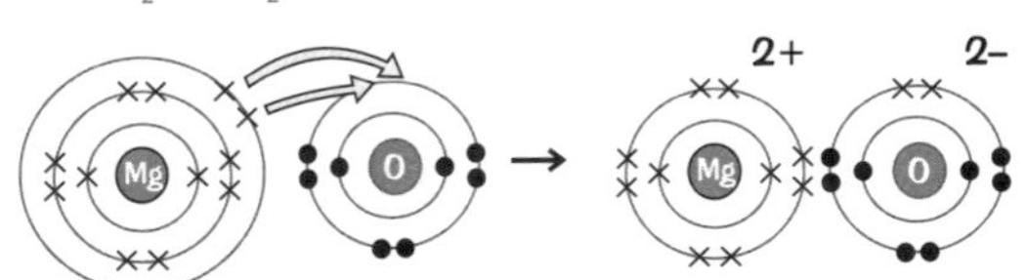

Section Six — Classifying Materials

b) A giant ionic lattice / structure.
c) strong, positive, negative, large

Q6a)

	Conducts electricity?
When solid	No
When dissolved in water	Yes
When molten	Yes

b) When solid, the ions are held tightly in a giant ionic structure so they're unable to move and conduct electricity. When dissolved or molten, the ions are free to move and so can conduct electricity.

Q7a)

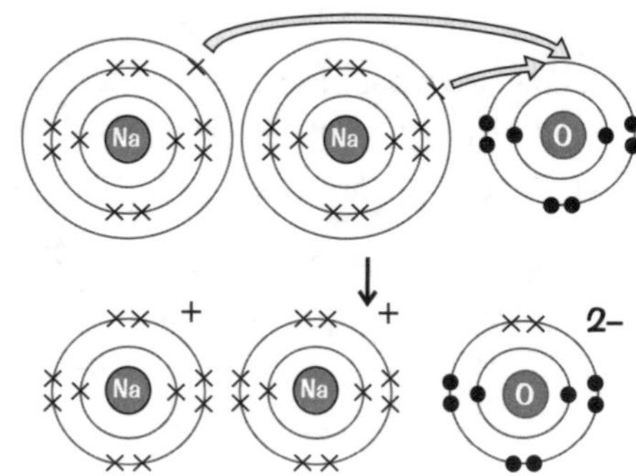

b)

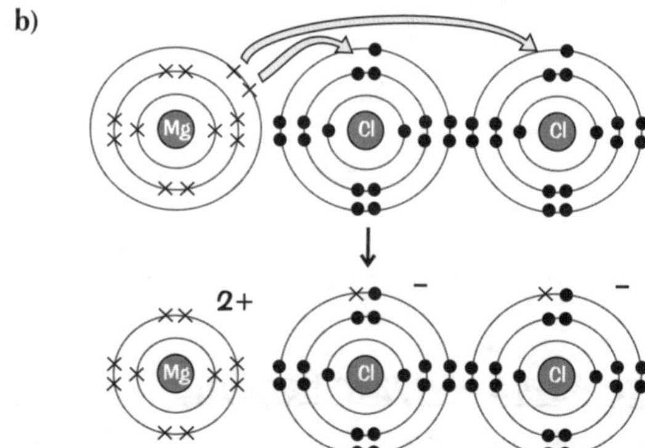

Page 47 — Ions and Formulas

Q1a) lose
b) gain
c) cations

Q2a) i) 1
ii) 2
iii) 1
b) i) 2
ii) 1
iii) 1

Q3a)

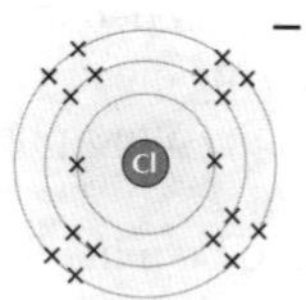

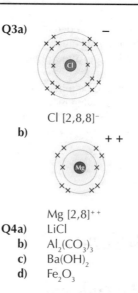

Cl $[2,8,8]^-$

b)

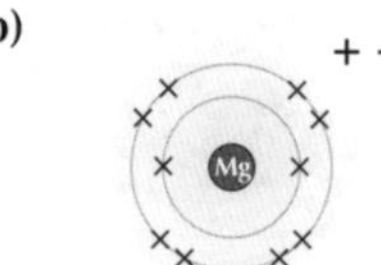

Mg $[2,8]^{++}$

Q4a) LiCl
b) $Al_2(CO_3)_3$
c) $Ba(OH)_2$
d) Fe_2O_3

Page 48 — Group I — Alkali Metals

Q1a) Lithium is less dense than water.
b) The solution would turn purple, because lithium hydroxide is formed which is a strong alkali.
c) $2Li + 2H_2O \rightarrow 2LiOH + H_2$
d) i) rubidium + water → rubidium hydroxide + hydrogen
ii) More vigorous — rubidium is more reactive as it's further down the group.

Q2a) potassium
b) sodium
c) lithium

Q3a) i) $Na \rightarrow Na^+ + e^-$
ii) It's oxidation because sodium loses an electron.
b) They both have one electron in their outer shell.
c) Potassium's outer electron is further away from the pull of the positive nucleus and is more shielded from it by inner shells, so it's more easily lost.

Page 49 — Group VII — Halogens

Q1a) i) $Cl_2 + 2e^- \rightarrow 2Cl^-$
ii) The process is reduction because the chlorine atoms are gaining electrons.
b) The outer shell in bromine atoms is further away and more shielded from the positive nucleus. This makes it less likely to pull in another electron to fill this shell.

Q2a) Bromine is more reactive than iodine so it displaces it from the potassium iodide solution. Bromine is less reactive than chlorine so it doesn't displace it from the solution.
b) $Br_2 + 2KI \rightarrow I_2 + 2KBr$
c) i) yes
ii) no

Q3a) sodium bromide
b) $2Na + Br_2 \rightarrow 2NaBr$
c) Faster, because bromine is more reactive than iodine.

Section Six — Classifying Materials

Page 50 — Covalent Bonding

Q1a) True
b) True
c) True
d) False
e) True
Q2a)

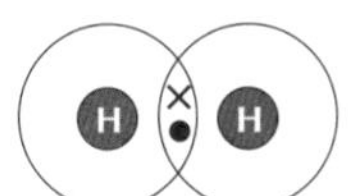

b)

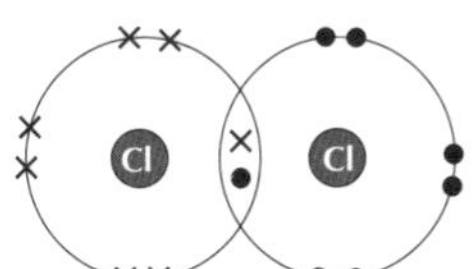

c)

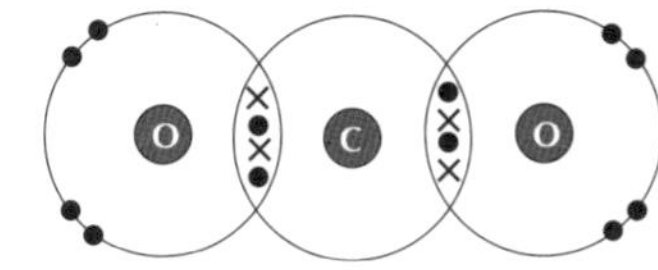

d)

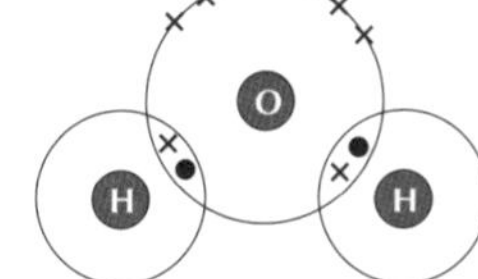

e)

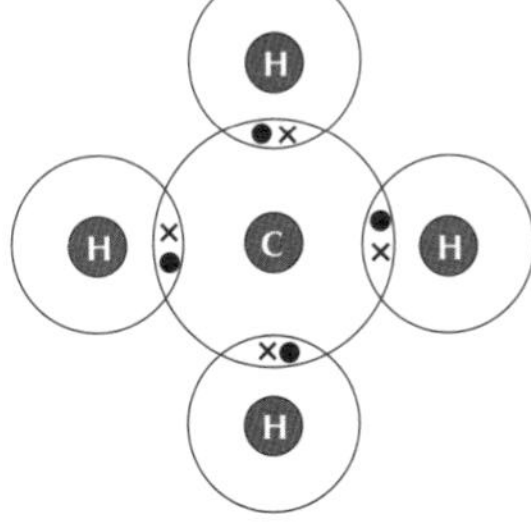

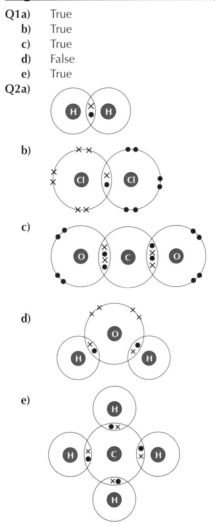

Q3 A chlorine atom only needs one electron to complete its outer shell, so it's very reactive. Two chlorine atoms share one electron each to form a single covalent bond. This gives both chlorine atoms a full outer shell of electrons which is a lot more stable.

Page 51 — Giant Covalent Structures

Q1 uncharged atoms, strong, high, insoluble
Q2a) Each carbon atom in graphite forms only three covalent bonds. Each has four outer electrons, so there are plenty of spare electrons not involved in bonds that are free to conduct electricity.
b) Each carbon atom in diamond forms four covalent bonds to give a very rigid structure held together very strongly. This makes it hard.
Q3a) Graphite — pencils — it is made of sheets of carbon atoms that are free to slide over each other. Layers can be rubbed off on to paper to give black pencil marks.
b) Diamond — glass-cutting tool — it is very hard.
Q4a) silicon dioxide
b) silicon and oxygen
c) glass

Page 52 — Simple Molecular Covalent Structures

Q1 small, strong, weak, easy
Q2 Any two of:
It will have a fairly low melting point.
It will have a fairly low boiling point.
It will be a gas or a liquid at room temperature.
It will not conduct electricity.
Q3a) low — the inter-molecular forces are weak and so the molecules can be easily parted from each other.
b) don't conduct — there are no free charges.
Q4a)

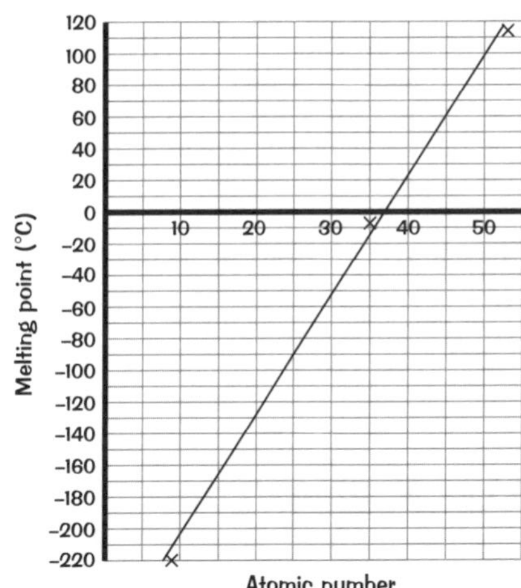

b) As the atomic number of the halogens increases, so do their melting points. The inter-molecular forces increase in strength with increasing atomic number (i.e. as the molecules get bigger). This means that it takes more energy to separate the molecules, so the melting point increases.

Page 53 — Group 0 — Noble Gases

Q1 The far-right column of the periodic table.
Q2a) colourless
b) low
c) neither gain nor lose
Q3a) The noble gases are colourless, have no smell and don't easily react with anything, so they're very difficult to detect.
b) Elements react in order to lose or gain enough electrons to give them a full outer shell. Noble gases already have a full outer shell of electrons so they do not need to react with anything.
c) Pass an electric current through them and they will give off a coloured glow.

Section Seven — Equations and Calculations

Q4 Neon — used in signs
Helium — used in airships and balloons
Argon — used in electric light bulbs

Q5a) Gas B is helium, because it glows yellow when electricity is passed through it / a balloon full of helium rises because it is less dense than air.

b) Gas C, because the brightly burning match shows that this gas isn't as unreactive as noble gases are.

Page 54 — Metallic Structures

Q1a) high
b) conductors
c) hammered, malleable

Q2 iron — ammonia production
nickel — converting natural oils into fats

Q3a) The metal heats up, which means some of the electrical energy is wasted as heat.

b) A superconductor is a material that has no electrical resistance. They can be made by cooling metals to very low temperatures.

c) E.g. power cables that transmit electricity without losing power, electromagnets that don't need a power source, very fast-working electronic circuits.

d) The temperatures that the materials have to be cooled to before they superconduct are extremely low, and therefore very expensive to produce.

Page 55 — Nanomaterials

Q1a) C_{60}
b) three
c) Yes, because (like graphite) it has spare electrons that are not involved in bonds. These electrons are therefore free to conduct electricity.

Q2a) In sunscreens — they reflect harmful UV radiation but not visible light, so you can't see them.

b) Silver nanoparticles are used in medicine to fight/kill viruses/microbes.

c) Any one of, e.g. in nanotubes to use in tiny electrical parts / to reinforce graphite in tennis racquets or building materials / industrial catalysts / highly specific sensors (e.g. for testing water purity).

Q3a) i) B
ii) Nanoparticles (1-100 nm) of titanium dioxide absorb visible light so would not make the paint appear white. Particles 10-100 mm are far too big and would make the paint lumpy.

b) absorb visible light — leaves no marks on the skin reflect UV light — prevents harmful rays reaching the skin
insoluble in water — not dissolved by sweat

Q4a) Nanomachines make only one specific product and can't replicate. They don't have the free will needed to change their activities and become destructive.

b) E.g. nanomachines could make products molecule by molecule, very quickly, cheaply and accurately and on a tiny scale.

Section Seven — Equations and Calculations

Page 56 — Relative Formula Mass

Q1a) How heavy an atom of the element is compared to an atom of carbon-12.

b) i) 24
ii) 20
iii) 16
iv) 1
v) 12
vi) 63.5
vii) 39
viii) 40
ix) 35.5

Q2 Element A is helium
Element B is (3 × 4) = 12 = carbon
Element C is (4 × 4) = 16 = oxygen

Q3a) (2 × 1) + 16 = **18**
b) 39 + 16 + 1 = **56**
c) 1 + 14 + (3 × 16) = **63**
d) (2 × 1) + 32 + (4 × 16) = **98**
e) 14 + (4 × 1) + 14 + (3 × 16) = **80**
f) (27 × 2) + 3[(32 + (16 × 4)] = **342**

Q4 $2XOH + H_2 = 114$
2 × (X + 16 + 1) + (2 × 1) = 114
2 × (X + 17) + 2 = 114
2 × (X + 17) = 112
X + 17 = 56
X = 39
so X = **potassium**

Page 57 — Isotopes and Relative Atomic Mass

Q1 isotopes, element, protons, neutrons

Q2 W and Y. These two atoms have the same number of protons (6) but different numbers of neutrons.

Q3a) i) 8
ii) 6
iii) 6

b) No — the chemistry of the two isotopes would be the same because they have the same number of electrons.

Section Seven — Equations and Calculations

Q4 Relative atomic mass is — the average mass of all atoms of that element.
Relative abundance means — the proportion of one isotope in an element.

Q5 $(35 \times 3) + (37 \times 1) = 142$
$142 \div 4 = \mathbf{35.5}$

Page 58 — Percentage Mass and Empirical Formulas

Q1a) Percentage mass of an element in a compound = $\frac{A_r \times \text{No. of atoms (of that element)}}{M_r \text{ (of whole compound)}} \times 100$

b) i) $(14 \times 2) \div [14 + (4 \times 1) + 14 + (3 \times 16)] \times 100 = 35\%$

ii) $(4 \times 1) \div [14 + (4 \times 1) + 14 + (3 \times 16)] \times 100 = 5\%$

iii) $(3 \times 16) \div [14 + (4 \times 1) + 14 + (3 \times 16)] \times 100 = 60\%$

Q2a) $14 \div (14 + 16) \times 100 = 46.7\%$ (to 3 s.f.)

b)

	Nitrogen	Oxygen
Percentage mass (%)	30.4	69.6
$\div A_r$	$(30.4 \div 14) = 2.17$	$(69.6 \div 16) = 4.35$
Ratio	1	2

empirical formula = NO_2

Q3

	Calcium	Oxygen	Hydrogen
Mass (g)	0.8	0.64	0.04
$\div A_r$	$(0.8 \div 40) = 0.02$	$(0.64 \div 16) = 0.04$	$(0.04 \div 1) = 0.04$
Ratio	1	2	2

empirical formula = $Ca(OH)_2$ (or CaO_2H_2)

Q4a) $A = (3 \times 16) \div [(2 \times 56) + (3 \times 16)] \times 100 = 30\%$
$B = 16 \div [(2 \times 1) + 16)] \times 100 = 89\%$ or 88.9%
$C = (3 \times 16) \div [40 + 12 + (3 \times 16)] \times 100 = 48\%$

b) B

Pages 59-60 — Calculating Masses in Reactions

Q1a) $2Mg + O_2 \rightarrow 2MgO$

b)
2Mg	2MgO
$2 \times 24 = 48$	$2 \times (24 + 16) = 80$
$48 \div 48 = 1$ g	$80 \div 48 = 1.67$ g
$1 \times 10 = 10$ g	$1.67 \times 10 = \mathbf{16.7\ g}$

Q2
4Na	$2Na_2O$
$4 \times 23 = 92$	$2 \times [(2 \times 23) + 16] = 124$
$92 \div 124 = 0.74$ g	$124 \div 124 = 1$ g
$0.74 \times 2 = \mathbf{1.48\ g}$	$1 \times 2 = 2$ g

Q3a) $2Al + Fe_2O_3 \rightarrow Al_2O_3 + 2Fe$

b)
Fe_2O_3	2Fe
$[(2 \times 56) + (3 \times 16)] = 160$	$2 \times 56 = 112$
$160 \div 160 = 1$ g	$112 \div 160 = 0.7$
$1 \times 20 = 20$ g	$0.7 \times 20 = \mathbf{14\ g}$

Q4 $CaCO_3 \rightarrow CaO + CO_2$

$CaCO_3$	CaO
$40 + 12 + (3 \times 16) = 100$	$40 + 16 = 56$
$100 \div 56 = 1.786$ kg	$56 \div 56 = 1$ kg
$1.786 \times 100 = \mathbf{178.6\ kg}$	$1 \times 100 = 100$ kg

Q5a)
C	2CO
12	$2 \times (12 + 16) = 56$
$12 \div 12 = 1$ g	$56 \div 12 = 4.67$ g
$1 \times 10 = 10$ g	$4.67 \times 10 = 46.7$ g

46.7 g of CO is produced in stage B — all this is used in stage C.

3CO	$3CO_2$
$3 \times (12 + 16) = 84$	$3 \times [12 + (16 \times 2)] = 132$
$84 \div 84 = 1$	$132 \div 84 = 1.57$
$1 \times 46.7 = 46.7$	$1.57 \times 46.7 = \mathbf{73.4\ g}$

b) It could be recycled and used in stage B.

Q6a) $2NaOH + H_2SO_4 \rightarrow Na_2SO_4 + 2H_2O$

b)
2NaOH	Na_2SO_4
$2 \times (23 + 16 + 1) = 80$	$(2 \times 23) + 32 + (4 \times 16) = 142$
$80 \div 142 = 0.56$ g	$142 \div 142 = 1$ g
$0.56 \times 75 = \mathbf{42\ g}$	$1 \times 75 = 75$ g

c)
H_2SO_4	$2H_2O$
$(2 \times 1) + 32 + (4 \times 16) = 98$	$2 \times [(2 \times 1) + 16] = 36$
$98 \div 98 = 1$ g	$36 \div 98 = 0.367$ g
$1 \times 50 = 50$ g	$0.367 \times 50 = \mathbf{18.4\ g}$

Pages 61-62 — The Mole

Q1a) mass, relative formula mass

b) i) 63.5 g

ii) $3 \times (2 \times 35.5) = \mathbf{213\ g}$

iii) $2 \times [1 + 14 + (16 \times 3)] = \mathbf{126\ g}$

iv) $0.5 \times [40 + 12 + (16 \times 3)] = \mathbf{50\ g}$

Q2a) number of moles = $\frac{\text{mass in g}}{M_r}$

b) i) $20 \div 40 = \mathbf{0.5\ moles}$

ii) $112 \div 32 = \mathbf{3.5\ moles}$

iii) $159 \div (63.5 + 16) = \mathbf{2\ moles}$

c) i) $2 \times 23 = \mathbf{46\ g}$

ii) $0.75 \times (24 + 16) = \mathbf{30\ g}$

iii) $0.025 \times [207 + (35.5 \times 2)] = \mathbf{6.95\ g}$

Q3a) moles = mass $\div A_r = 13 \div 65 = \mathbf{0.2\ moles}$

b) $2 \times 0.2 = \mathbf{0.4\ moles}$

Q4a) no. of moles = concentration (mol/dm^3) $\times$ volume (dm^3)

b) i) $0.05 \times 2 = \mathbf{0.1\ mol}$

ii) $0.25 \times 0.5 = \mathbf{0.125\ mol}$

iii) $0.55 \times 1.75 = \mathbf{0.9625\ mol}$

Q5a) $0.05 \div 0.5 = \mathbf{0.1\ mol/dm^3}$

b) $0.2 \div 0.25 = \mathbf{0.8\ mol/dm^3}$

c) $0.03 \div 0.02 = \mathbf{1.5\ mol/dm^3}$

Q6a) $2Na + 2H_2O \rightarrow 2NaOH + H_2$

b) Moles of Na = $0.6 \div 23 = 0.026$
Half this number of moles of H_2 were produced.
So moles of H_2 produced = $0.026 \div 2 = 0.013$
mass of H_2 produced = moles $\times M_r$
$= 0.013 \times 2 = \mathbf{0.026\ g}$

c) Number of moles of NaOH formed = 0.026
mass of NaOH produced =
0.06 × (23 + 16 + 1) = **1.04 g**

Page 63 — Atom Economy

Q1a) copper
b) M_r of CuO = 63.5 + 16 = 79.5
(2 × 63.5) ÷ [(2 × 79.5) + 12] × 100 = **74.3%**
c) 100 – 74.3 = **25.7%**
Q2a) Reactions with a high atom economy use up resources less quickly and produce less waste. This reduces the cost of disposing of the waste safely, and of buying more raw resources. This all means more profit.
b) Reactions that only have one product, e.g. the Haber process.
Q3a) M_r of $TiCl_4$ = 48 + (4 × 35.5) = 190
i) With magnesium: 48 ÷ [190 + (2 × 24)] × 100 = **20.2%**
ii) With sodium: 48 ÷ [190 + (4 × 23)] × 100 = **17%**
b) The method with magnesium has the best atom economy.
Q4 $Cr_2O_3 + 2Al \rightarrow Al_2O_3 + 2Cr$
M_r of Cr_2O_3 = (2 × 52) + (3 × 16) = 152
(2 × 52) ÷ [152 + (2 × 27)] × 100 = **50.5%**

Page 64 — Percentage Yield

Q1a) Percentage yield
= (actual yield ÷ theoretical yield) × 100
b) (1.2 ÷ 2.7) × 100 = **44.4%**
Q2a) (6 ÷ 15) × 100 = **40%**
b) When the original reactants were mixed, a small amount will have been left in their containers. When the solid was filtered out, some of it will have been left on the piece of filter paper. (Also accept: there may have been unexpected reactions due to impurities / changes in the reaction conditions / a mistake may have been made in the original calculation / the reactants may not have been measured out accurately.)
Q3a) A = (3.00 ÷ 3.33) × 100 = **90.1%**
B = (3.18 ÷ 3.33) × 100 = **95.5%**
C = (3.05 ÷ 3.33) × 100 = **91.6%**
D = (3.15 ÷ 3.33) × 100 = **94.6%**
b) C, D

Section Eight — Chemical Change

Pages 65-66 — Acids and Bases

Q1a) acid + base → **salt** + **water**
b) neutralisation
c) i) $H^+(aq)$ and $OH^-(aq)$
ii) $H^+(aq)$
iii) $OH^-(aq)$
iv) $Na^+(aq)$ and $Cl^-(aq)$
v) $OH^-(aq)$
vi) $H^+(aq)$
Q2a) i)

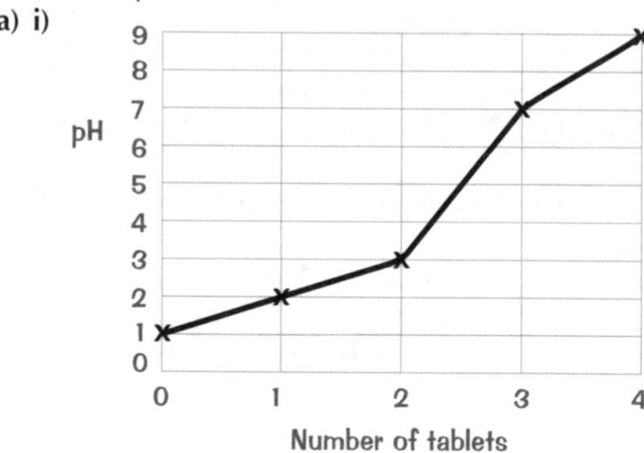

ii) The pH increases from pH 1 to pH 9. It increases most sharply between pH 3 and pH 7.
iii) 3
b) E.g. the new line should be above the previous one and reach pH 7 at 2 tablets.

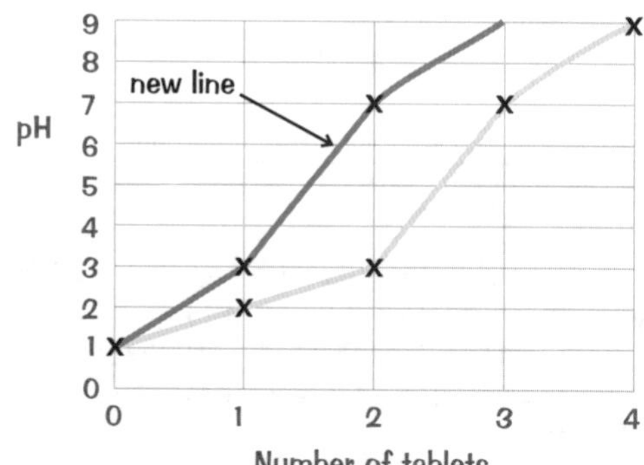

Q3a) neutral
b) dyes
c) 7, green
Q4a) Test the solution with universal indicator or with a pH meter.
b) Baking soda or soap powder, because they are weak bases and so would neutralise the acid but wouldn't irritate or harm the skin. Stronger bases like caustic soda might damage the skin.
c) The colour can be difficult to judge exactly / Dyes can't give an exact reading, only a general idea / other answers are possible.
Q5a) E.g. car batteries; cleaning metal surfaces; paints
b) ammonium sulfate
c) Sulfuric acid's used to clean the metal surface. It reacts with insoluble metal oxides, to form soluble metal salts which wash away easily.

Section Eight — Chemical Change

Pages 67-68 — Acids Reacting with Metals

Q1 metals, hydrogen, silver, reactive, more, chlorides, sulfuric, nitric.

Q2a) E.g. the number of gas bubbles produced in a certain time / the time it takes for the metal to disappear completely / the volume of gas produced in a certain time / loss of mass in a certain time.

b) acid concentration

c) Two from: volume of acid, the mass / size of the metal pieces, temperature.

Q3a) i) $Ca + \mathbf{2}HCl \rightarrow CaCl_2 + H_2$

ii) $\mathbf{2}Na + \mathbf{2}HCl \rightarrow \mathbf{2}NaCl + H_2$

iii) $\mathbf{2}Li + H_2SO_4 \rightarrow Li_2SO_4 + H_2$

b) i) magnesium bromide

ii) $2Al + 6HBr \rightarrow 2AlBr_3 + 3H_2$

Q4a)

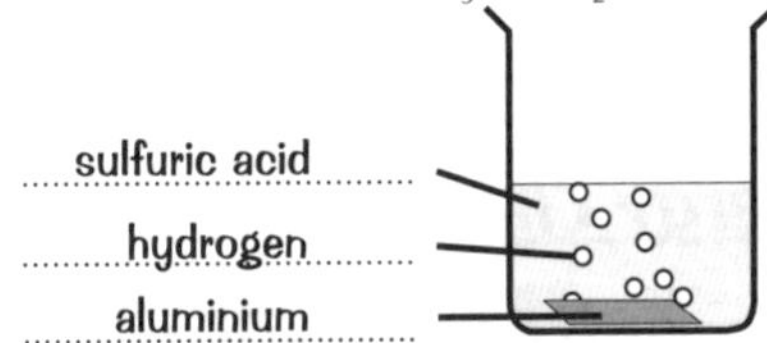

b) aluminium + **sulfuric acid** → aluminium sulfate + **hydrogen**

c) $\mathbf{2Al + 3H_2SO_4 \rightarrow Al_2(SO_4)_3 + 3H_2}$

d) zinc + sulfuric acid → zinc sulfate + hydrogen

e) i) $Mg + 2HCl \rightarrow MgCl_2 + H_2$

ii) $Ca + 2HNO_3 \rightarrow Ca(NO_3)_2 + H_2$

Q5a) A

b) B

c) A: magnesium
B: copper
C: iron
D: zinc

Pages 69-70 — Neutralisation Reactions

Q1a) hydrochloric acid + lead oxide → **lead** chloride + water

b) nitric acid + copper hydroxide → copper **nitrate** + water

c) sulfuric acid + zinc oxide → zinc sulfate + **water**

d) hydrochloric acid + **nickel** oxide → nickel **chloride** + **water**

e) **nitric** acid + copper oxide → **copper** nitrate + **water**

f) sulfuric acid + **sodium** hydroxide → sodium **sulfate** + **water**

Q2a) The following should be ticked:
Acids react with metal oxides to form a salt and water.
Salts and water are formed when acids react with metal hydroxides.
Ammonia solution is alkaline.

b) $H_2SO_4 + CuO \rightarrow CuSO_4 + H_2O$
$HCl + NaOH \rightarrow NaCl + H_2O$

Q3a) E.g. potassium oxide / hydroxide + sulfuric acid

b) ammonia + hydrochloric acid

c) E.g. silver oxide / hydroxide + nitric acid

Q4a) NH_3

b) gas, alkaline, nitrogen, proteins, salts, fertilisers

c) ammonia + nitric acid → ammonium nitrate

d) Because it has nitrogen from two sources, the ammonia and the nitric acid.

e) No water is produced.

Q5a) i) $CuO(s)$

ii) $H_2O(l)$

iii) $HCl(aq)$

iv) $ZnO(aq)$

v) $Na_2SO_4(aq) + H_2O(l)$

b) i) $\mathbf{2}NaOH + H_2SO_4 \rightarrow Na_2SO_4 + \mathbf{2}H_2O$

ii) $Mg(OH)_2 + \mathbf{2}HNO_3 \rightarrow Mg(NO_3)_2 + \mathbf{2}H_2O$

iii) $\mathbf{2}NH_3 + H_2SO_4 \rightarrow (NH_4)_2SO_4$

Pages 71-72 — Making Salts

Q1a) soluble

b) insoluble

c) insoluble, neutralised

d) precipitation

e) more, less

Q2a) B

b) D

c) A

d) C

Q3a) The magnesium displaces the copper from the solution, and the displaced copper coats the piece of magnesium.

b) magnesium sulfate

c) The magnesium isn't in contact with the solution any more, as it is covered with a layer of copper.

d) displacement

Q4a) **silver nitrate** + **sodium chloride** → silver chloride + **sodium nitrate**

b) The silver chloride must be filtered out of the solution. It needs to be washed and then dried on filter paper / by heating.

Q5a) a spatula

b) The nickel carbonate no longer dissolves / solid remains in the flask and bubbles no longer form.

c)

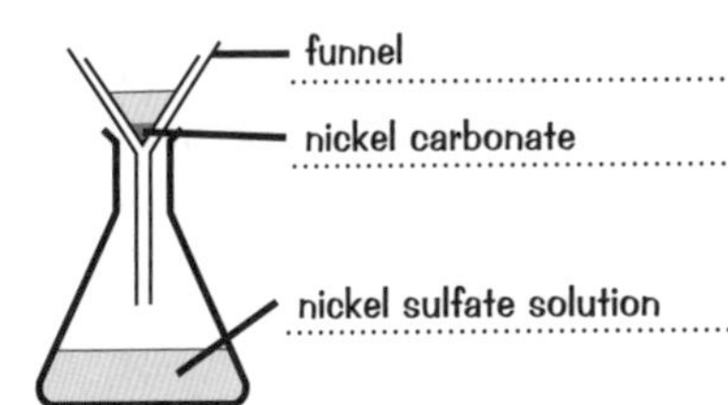

d) filtration

e) Heat the nickel sulfate solution to evaporate off the water.

Section Nine — Reaction Rates and Energy Changes

Page 73-74 — More Chemical Changes

Q1a) $Cu(OH)_2$ (s) → CuO (s) + H_2O (l)
ethanol + ethanoic acid
→ ethyl ethanoate + water

b) CaO (s) + H_2O (l) → $Ca(OH)_2$ (s)

Q2a) i) False
ii) True
iii) True
iv) False

b) sodium hydrogencarbonate → sodium carbonate + carbon dioxide + water

Q3a) Because water is removed from the hydrated copper sulfate in the reaction.

b) Because the hydrated copper sulfate breaks down into two simpler substances on heating.

c) Water reacts with the anhydrous copper sulfate to form a new substance.

Q4a) precipitation

b) copper hydroxide

c) $CuSO_4 + 2NaOH \rightarrow Cu(OH)_2 + Na_2SO_4$

Q5a)

Compound	Metal Cation	Colour of Precipitate
copper(II) sulfate	**Cu^{2+}**	blue
iron(II) sulfate	**Fe^{2+}**	**dark grey/green**
iron(III) chloride	Fe^{3+}	**orange**
copper(II) chloride	**Cu^{2+}**	**blue**

b) $CuCl_2 + 2NaOH \rightarrow Cu(OH)_2 + 2NaCl$

Q6a) B

b) water and oxygen

c) iron + oxygen + water → hydrated iron(III) oxide

d) Because salty water / sea spray accelerates rusting.

Page 75-76 — Electrolysis and the Half-Equations

Q1 electricity, electrolyte, electrons, anode, cathode

Q2a) A: H^+
B: Cl^-
C: H_2
D: Cl_2

b) Cathode: $2H^+ + 2e^- \rightarrow H_2$
Anode: $2Cl^- \rightarrow Cl_2 + 2e^-$

Q3 Electrolysis is the breaking down of substances using electricity. For electricity to flow, charged ions need to be free to move, and so you need a liquid.

Q4 Cathode: $Cu^{2+}(aq) + 2e^- \rightarrow Cu(s)$
Anode: $Cu(s) \rightarrow Cu^{2+}(aq) + 2e^-$

Q5 The impurities are not charged (i.e. they are neutral) so they are not attracted to the cathode.

Q6a) Copper is often used in electrical conductors, and the purer it is, the better it conducts.

b) impure copper

c) The electrical supply pulls electrons off the atoms of copper at the anode, giving Cu^{2+} ions. These positive ions are attracted to the negative cathode, where the ions get electrons back to make them pure copper atoms, which bind to the cathode.

Q7 Cathode: $Ag^+(aq) + e^- \rightarrow Ag(s)$
Anode: $Ag(s) \rightarrow Ag^+(aq) + e^-$

Q8 The copper produced will have zinc impurities in it.

Section Nine — Reaction Rates and Energy Changes

Page 77 — Rates of Reaction

Q1a) faster

b) increase

c) gases, faster

d) decreases

e) speeds up, isn't

Q2a) i) Z

ii) The gas produced by the reaction was given off more quickly, shown by the steeper curve. / The reaction finished more quickly, shown by the curve levelling off sooner.

b) Because the same mass of marble (and the same amount of acid) reacted each time.

c) The curve should have the same initial slope as curve Z, but show that a larger volume of gas is produced, e.g. like this:

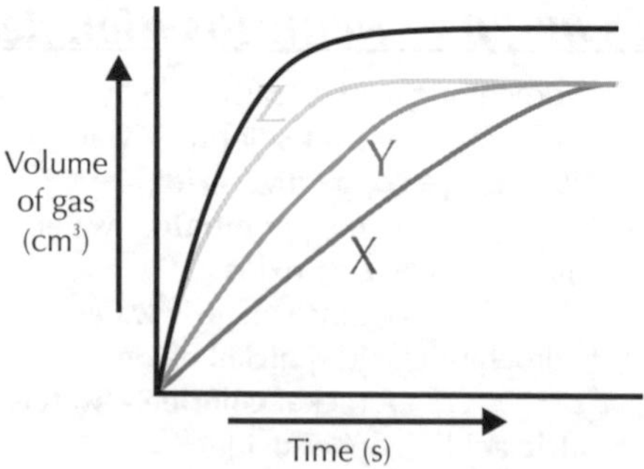

Q3a) More reactants were used and so more products (including the gas) were formed.

b) Reaction Q might have been carried out at a higher temperature / one of the reactants may have been more concentrated / a catalyst might have been used / a solid reactant might have had a larger surface area.

Section Nine — Reaction Rates and Energy Changes

Pages 78-79 — Measuring Rates of Reaction

Q1a) speed, reactants, formed
b) precipitation, faster
c) gas, mass, volume
Q2a) i) $0.75 \div 5 = \mathbf{0.15\ g/s}$
ii) $0.25 \div 5 = \mathbf{0.05\ g/s}$
b) C
Q3a)

Concentration of HCl (mol/dm³)	Experiment 1 — volume of gas produced (cm³)	Experiment 2 — volume of gas produced (cm³)	Average volume of gas produced (cm³)
2	92	96	94
1.5	63	65	64
1	44	47	45.5
0.5	20	50	35
0.25	9	9	9

b) 50 in the third column of the table should be circled.
c) 2 mol/dm³
d) i) gas syringe
ii) Stopwatch / stopclock / timer / balance / measuring cylinder
e) Sketch should look something like this:

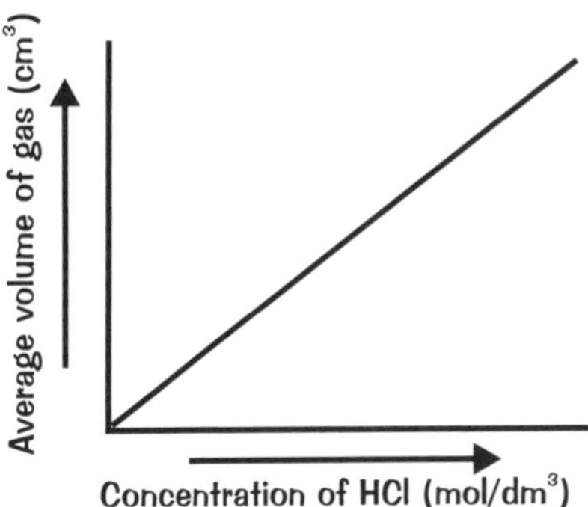

f) To improve the reliability of the results.
g) Bad experimental technique, e.g. misreading value from gas syringe, not emptying gas syringe before starting, using the wrong concentration of acid.
h) Any two reasonable suggestions, e.g. take more readings for each concentration, use more concentrations.

Page 80 — Collision Theory

Q1a) collide
b) energy
c) faster, more
d) rate
Q2 increasing the temperature — makes the particles move faster so they collide more often
decreasing the concentration — means fewer particles of reactant are present, so fewer collisions occur
adding a catalyst — provides a surface for particles to stick to and lowers activation energy
increasing the surface area — gives particles a bigger area of solid reactant to react with

Q3a) i) increase
ii) Increasing the pressure makes the particles more squashed up together, so there is more chance of them colliding and reacting.
b)

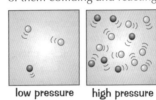

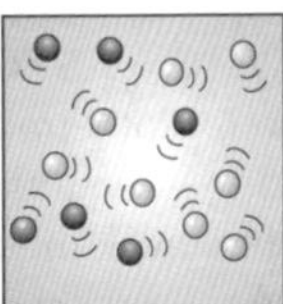

Q4a) False
b) True
c) False
d) True
Q5 A catalyst changes the rate of reaction without being changed or used up in the reaction. It increases the number of successful collisions by lowering the activation energy.

Page 81 — Catalysts

Q1a) activation energy
b) A
c)

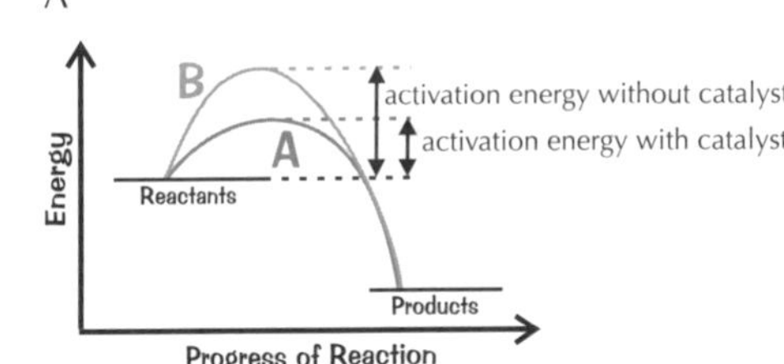

Q2 To give the catalysts a large surface area — the greater the surface area of catalyst exposed, the better it works.
Q3a) transition metals
b) E.g. iron is used in the Haber process. (Other correct answers are possible, e.g. nickel is used in hydrogenation of vegetable oil to make margarine.)
Q4a) They increase the rate of reaction, so the plant doesn't need to operate for as long to give the same amount of product.
They allow the reaction to happen at a lower temperature, so energy costs are lower.
b) Any two of:
They can be very expensive to buy.
You might have to buy several different catalysts if your plant uses several different reactions.
The catalyst often needs to be removed from the product and cleaned.
Catalysts can be poisoned by impurities and stop working.

Section Nine — Reaction Rates and Energy Changes

Pages 82-83 — Energy Transfer in Reactions

Q1 energy, exothermic, heat, an increase, endothermic, heat, a decrease

Q2a) 29.5 °C – 22 °C = **7.5 °C** (accept 7 °C)

b) neutralisation, exothermic

Q3 Bond breaking — endothermic — energy must be supplied to break the bonds.
Bond forming — exothermic — energy is released when new bonds are formed.

Q4a) 1800 000 ÷ 1000 = 1800 kJ

b) 1800 000 ÷ 90 000 = 20
1000 kg ÷ 20 = 50 kg

Q5a) Energy is taken in.

b) Energy is given out.

c) Burning methane is an exothermic process (this is true of all fuels) — when methane burns it gives out heat.

d) B

Q6a) X

b) N

c) X

d) N

Page 84-85 — Bond Energies

Q1a) A, C and D

b) B

c) B and E

d) D

Q2a) the energy change / the enthalpy change

b) endothermic

c) exothermic

d) The energy that is initially needed to break the bonds in the reactants.

e) They lower the activation energy.

Q3a) $\Delta H = 120 - 30 = -90$ kJ/mol

b) Activation energy = 190 – 120 = +70 kJ/mol

c) Activation energy = 90 + 70 = +160 kJ/mol

Q4a) (4 × 412) + (2 × 498) = 2644 kJ/mol

b) (2 × 743) + (4 × 463) = 3338 kJ/mol

c) $\Delta H = 2644 - 3338 = -694$ kJ/mol

Q5 $\Delta H = [158 + (4 \times 391) + 498] - [945 + (4 \times 463)] = -577$ kJ/mol

Q6a) $\Delta H = [(2 \times 348) + (12 \times 412) + (7 \times 498)] - [(8 \times 743) + (12 \times 463)] = -2374$ kJ/mol

b) $\Delta H = -2374 \div 2 = -1187$ kJ/mol

Page 86 — Measuring the Energy Content of Fuels

Q1a) energy transferred = mass of water × specific heat capacity of water (4.2) × temperature change

b) energy given out per gram = energy transferred ÷ mass of fuel burned

Q2a)

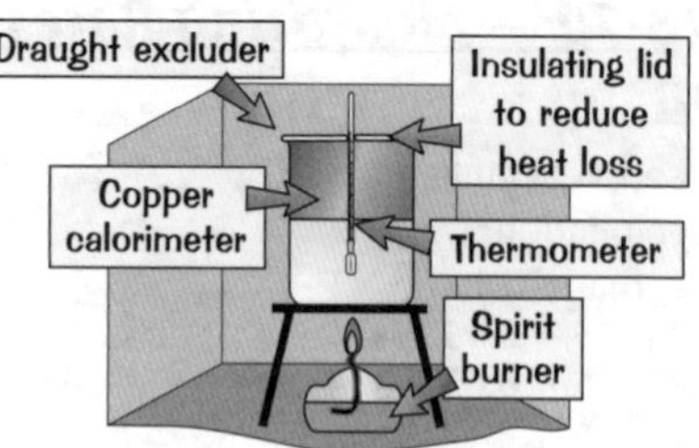

b) E.g. any two of: volume of each fuel / same apparatus — use of lid, material of can, use of draught excluder / mass of water used / distance of spirit burner from can.

c) i) Energy gain = 50 × 4.2 × 30.5 = 6405 J

ii) Energy per gram = 6405 ÷ 0.7 = 9150 J/g = 9.15 kJ/g

d) Energy gain = 50 × 4.2 × 27 = 5670 J
Energy produced per gram = 5670 ÷ 0.8 = 7087.5 J/g = 7.09 kJ/g

e) Petrol would make the better fuel because it releases more energy per gram than fuel X does.

Page 87 — Reversible Reactions

Q1a) products, react, reactants

b) balance

c) closed, escape

Q2a) True

b) False

c) True

d) False

e) False

Q3a) temperature, pressure

b) It takes in heat — all reversible reactions are exothermic in one direction and endothermic in the other.

c) backward

d) Because one reaction is always exothermic and the other endothermic.

e) It won't affect the position, as the volume (number of molecules) of products and reactants are the same.

Page 88 — The Haber Process

Q1a) nitrogen and hydrogen

b) The left (reactants) side.

c) The pressure should be increased, as this will encourage the reaction which produces less volume, i.e. the one which produces more ammonia.

Q2a) 200 atmospheres, 450 °C.

b) The pressure should be as high as possible to encourage the forward reaction.
High pressures are expensive to create, so the pressure must not be so high that the costs outweigh the benefits.

Section Ten — Chemical Tests

Q3a) Raising the temperature will reduce the amount of ammonia formed.

b) Because if a low temperature was used the reaction would be far too slow.

c) They are recycled and used to produce more product.

c) NH_4^+ (aq) + OH^- (aq) → NH_3 (g) + H_2O (l)

Q5a) $CuSO_4$

b) $Al_2(SO_4)_3$

c) $FeSO_4$

d) $FeCl_3$

e) NH_4Cl

f) $CaCl_2$

Section Ten — Chemical Tests

Page 89 — Gas Tests

Q1a) B

b) C and D

c) C

d) A (only a small amount of CO_2 dissolves) and D

e) D

Q2a) chlorine

b) hydrogen

c) ammonia

d) oxygen

Q3a)

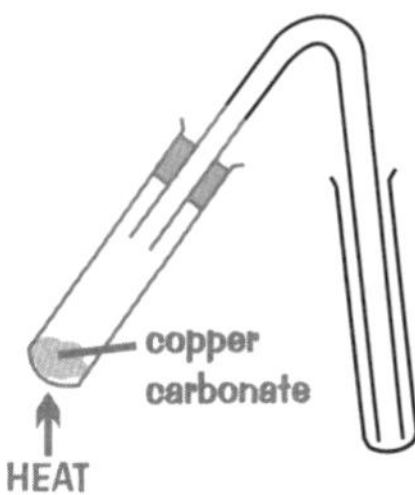

b) Because it is heavier than air.

c) Test with damp litmus paper (for chlorine). Test with a glowing splint (for oxygen).

Pages 90-91 — Tests for Positive Ions

Q1a) A white precipitate forms in the solution.

b) $CaCl_2$ (aq) + 2NaOH (aq) → $Ca(OH)_2$ (s) + 2NaCl (aq)

c) Ca^{2+} (aq) + $2OH^-$ (aq) → $Ca(OH)_2$ (s)

Q2a) brick-red flame — Ca^{2+}
yellow/orange flame — Na^+
blue-green flame — Cu^{2+}
lilac flame — K^+

b) potassium nitrate

Q3a) Fe^{2+} **(aq)** + **2**OH^- (aq) → $\mathbf{Fe(OH)_2}$ (s)

b) Fe^{3+} (aq) + $3OH^-$ (aq) → $Fe(OH)_3$ (s)

c) Amy would see a white precipitate at first, but it would re-dissolve in excess NaOH to form a colourless solution.

Q4a) Warm the mixture, testing any gas given off with a damp piece of universal indicator or litmus paper.

b) A pungent-smelling gas is given off when the solid is warmed. This gas turns the damp universal indicator paper purple or the litmus paper blue.

Page 92 — Tests for Negative Ions

Q1a) SO_4^{2-}

b) I^-

c) Br^-

Q2 hydrochloric acid, carbon dioxide, limewater

Q3a) dilute hydrochloric acid, barium chloride

b) a white precipitate (of barium sulfate)

c) i) dilute hydrochloric acid

ii) Sulfur dioxide will be produced which turns damp potassium dichromate(VI) paper from orange to green.

Q4 By adding dilute nitric acid to a solution of the compound, and then adding some silver nitrate solution. If the compound contains Cl^- ions a white precipitate will form. If it contains Br^- ions a cream precipitate will form. If it contains I^- ions a yellow precipitate will form.

Q5a) Ag^+ (aq) + $\mathbf{Cl^-}$ **(aq)** → AgCl (s)

b) 2HCl (aq) + Na_2CO_3 (s) → 2NaCl (aq) + $\mathbf{H_2O}$ (l) + $\mathbf{CO_2}$ (g)

c) $\mathbf{Ba^{2+}}$ **(aq)** + $\mathbf{SO_4^{2-}}$ **(aq)** → $BaSO_4$ (s)

Page 93 — Tests for Acids and Alkalis

Q1a) blue, red, red, blue

b) H^+ ions

c) Add the substance to a fairly reactive metal (e.g. magnesium). Hydrogen will be given off in the presence of H^+ ions, which can be tested for using the 'squeaky pop' test. / Add the substance to a carbonate (e.g. calcium carbonate). Carbon dioxide will be given off in the presence of H^+ ions, which can be tested for using limewater.

d) OH^- ions

e) By heating the substance with an ammonium salt. Ammonia gas, which has a pungent smell, will be given off if OH^- ions are present.

Q2a) $Ca(OH)_2 + 2NH_4Cl \rightarrow CaCl_2 + 2NH_3 + 2H_2O$

b) $NH_4^+ + OH^- \rightarrow NH_3 + H_2O$

c) Ammonia turns damp universal indicator paper purple / damp red litmus paper blue.

d) pink

Section Eleven — Water and Equilibria

Pages 94-95 — Tests for Organic Compounds

Q1a) yellowy-orange and/or blue
b) Octane will be smokier. It has a higher proportion of carbon.
Q2a) i) The volume of organic compound used, the volume of bromine water used, the time for which each sample is shaken.
ii) To make it a fair test.
b) B
c) A
d) The last two (with double or triple bonds between the carbon atoms).
Q3a) $(12 \div 44) \times 8.8 = 2.4$ g of carbon
b) $(2 \div 18) \times 5.4 = 0.6$ g of hydrogen
c) moles of C = $2.4 \div 12 = 0.2$ mol
moles of H = $0.6 \div 1 = 0.6$ mol
d) $0.2 : 0.6 = (0.2 \div 0.2) : (0.6 \div 0.2) = 1 : 3$ ratio
empirical formula is $\mathbf{CH_3}$
Q4 mass of C = $(12 \div 44) \times 4.4 = 1.2$ g
mass of H = $(2 \div 18) \times 1.8 = 0.2$ g
moles of C = $1.2 \div 12 = 0.1$ mol
moles of H = $0.2 \div 1 = 0.2$ mol
ratio = $0.1 : 0.2 = (0.1 \div 0.1) : (0.2 \div 0.1) = 1 : 2$
empirical formula = $\mathbf{CH_2}$
Q5 mass of C = $(12 \div 44) \times 1.1 = 0.3$ g
mass of H = $(2 \div 18) \times 0.9 = 0.1$ g
mass of O = $0.8 - 0.3 - 0.1 = 0.4$ g
moles of C = $0.3 \div 12 = 0.025$ mol
moles of H = $0.1 \div 1 = 0.1$ mol
moles of O = $0.4 \div 16 = 0.025$ mol
ratio = $0.025 : 0.1 : 0.025 = 1 : 4 : 1$
empirical formula = $\mathbf{CH_4O}$

Page 96 — Instrumental Methods

Q1a) They are much faster than lab methods.
b) Any two of: They are more accurate than lab methods / They are sensitive, and can be used even with a very small sample / They don't need trained chemists to carry out the tests / The tests can be automated.
Q2a) lithium, Li
b) E.g. the steel industry.
Q3a) chromium, manganese and iron
b) C_5H_{10}
c) Antimony has two isotopes, one with an atomic mass of 121, and one with an atomic mass of 123.
Q4a) nuclear magnetic resonance
b) organic compounds
c) hydrogen atoms

Section Eleven — Water and Equilibria

Pages 97-98 — Water

Q1a) air rises, water condenses — B
water flows — D
evaporation — A
rain — C
b) A
c) D
Q2a)

SALT	SOLUBLE	INSOLUBLE
sodium sulfate	✓	
ammonium chloride	✓	
lead nitrate	✓	
silver chloride		✓
lead sulfate		✓
potassium chloride	✓	
barium sulfate		✓

b) i) Potassium chromate — soluble because all salts of potassium are soluble.
ii) Vanadium nitrate — soluble because all nitrates are soluble.
Q3 pure, sulfur, power stations, acid, fertilisers, rocks, households
Q4 The following statements should be ticked:
Water is essential for life because many biological reactions take place in solution.
Fertilisers are often ammonium salts which are soluble in water.
Many covalent compounds, like wax and petrol, don't dissolve in water.
Q5a)

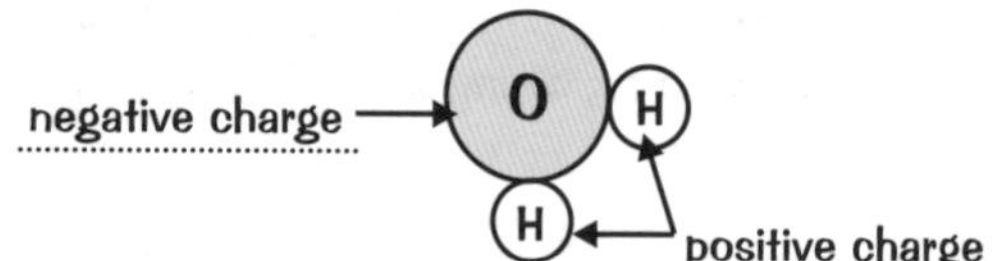

b) i)

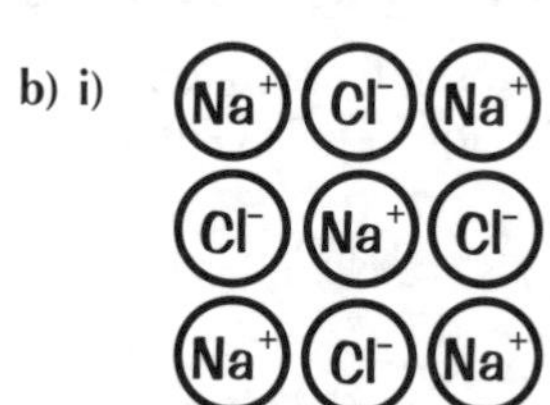

part of an NaCl crystal

ii) Ionic bonds / the attraction of the opposite charges / electrostatic forces of attraction.
iii) Diagram should look similar to:

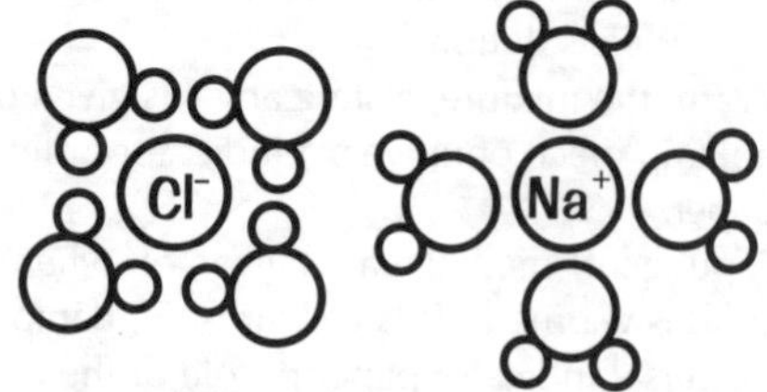

dissolved sodium and chloride ions

Section Eleven — Water and Equilibria

Pages 99-100 — Hard Water

Q1a) True
b) False
c) False
d) False
e) True
Q2a) Tap water might already contain dissolved ions, which could affect the results.
b) i) magnesium sulfate and calcium chloride
ii) Because more soap was needed to produce a lather with these compounds / a scum was formed when soap was added.
c) The distilled water sample acted as the control.
d) It works well in both hard and soft water.
Q3a) Scale coats the heating element and so the kettle becomes less efficient / takes longer to boil.
b) The scale build-up means less water can flow, and it could eventually block pipes altogether.
Q4 $Na_2Resin(s) + Ca^{2+}(aq) \rightarrow CaResin(s) + 2Na^+(aq)$ / $Na_2Resin(s) + Mg^{2+}(aq) \rightarrow MgResin(s) + 2Na^+(aq)$
The resin has sodium ions attached to it. It exchanges these for the Ca^{2+} and Mg^{2+} ions that make the water hard, removing them from the water.
Q5a) Calcium carbonate isn't soluble in water, so there were no Ca^{2+} ions in solution.
b) i) $CO_2(g) + H_2O(l) + CaCO_3(s) \rightarrow \mathbf{Ca(HCO_3)_2}(aq)$
ii) The $Ca(HCO_3)_2$ is soluble in water, so there are now calcium ions in solution.
c) i) calcium carbonate (also accept 'scale')
ii) calcium hydrogencarbonate → calcium carbonate + water + carbon dioxide
d) Using an acid — calcium carbonate dissolves in acid.

Page 101 — Water Quality

Q1a) distilled water
b) chlorine taste
Q2 filtration — solids
chlorination — harmful microorganisms
adding iron compounds — phosphates
adding lime (calcium hydroxide) — acidity
Q3a) B
b) D
c) A
d) C
Q4a) It's too expensive, and people often live in isolated rural areas where the nearest source of water is a long distance away / the countries don't have the infrastructure to supply clean water.
b) The biggest increases in life expectancy have been related to the ability of countries to supply clean water.
Q5a) They are poisonous.
b) bacteria / chemicals
c) by filtration (using gravel beds)

Page 102 — Solubility

Q1a) grams per 100 grams of solvent
b) no more solute dissolves at that temperature
c) low, high
Q2 aquatic, oxygen, chlorine, bleach, carbon dioxide, pressure, fizzy
Q3a) 55 g / 100 g
b) 105 g / 100 g (accept any answer between 103 and 105 g / 100 g)
c) i) lead nitrate
ii) potassium nitrate
d) 85 – 38 = 47 g (accept any answer between 45 and 50 g)

Page 103-104 — Detergents and Dry-Cleaning

Q1a) hydrophilic → [head]〰〰 ← hydrophobic
b) i) the hydrophilic head
ii) the hydrophobic tail
c) i) they form scum
ii) crude oil
Q2a) Biological detergents contain enzymes.
b) blood stains, grass stains, tomato ketchup stains
Q3a) Washing powder A
b) Washing powders A and E — they're more effective at lower temperatures.
Q4a) High temperatures help to melt and disperse greasy dirt deposits.
b) i) Wool shrinks at high temperatures.
ii) Nylon garments can lose their shape at high temperatures.
c) E.g. More energy is needed to give the high temperature and so more carbon dioxide is emitted. This contributes to the greenhouse effect/climate change.
Q5 intermolecular, bonds, surrounded, detergents, solvent
Q6a) 100 g of solvent A dissolves 12.1 g paint
100 ÷ 12.1 = 8.26 g solvent A dissolves 1 g paint
8.26 × 50 = **413 g** solvent A dissolves 50 g paint
b) Solvent A — it dissolves more paint than the other solvents so is likely to form stronger intermolecular bonds.
c) Solvent C — it dissolves paint nearly as well as solvent A, but it's a lot cheaper.

Page 105 — Changing Equilibrium

Q1a) C
b) A
c) B
Q2a) It increases.
b) It decreases.
c) As the temperature increases the amount of ammonia decreases, so this must be an exothermic reaction — the equilibrium is attempting to counteract the change.

Section Twelve — Concentrations and Electrolysis

Q3a) 2
b) the amount decreases
c) The percentage of ethanol decreases because the equilibrium shifts to the left to oppose the change — there are more molecules of gas on the left-hand side of the equation.

Page 106 — Acid-Base Theories

Q1a) $H^+(aq)$
b) $OH^-(aq)$
c) $H^+(aq)$
Q2a) i) When mixed with **water**, all acids release **H^+ / hydrogen** ions.
ii) When mixed with **water**, all alkalis form **OH^- / hydroxide** ions.
b) Ammonia gas behaved as a base even when it wasn't dissolved in water / ammonia gas does not contain OH^- ions.
c) Charged subatomic particles had not been discovered in the 1880s / scientists didn't realise it was possible to have charged ions at the time.
Q3a) H^+
b) Lowry and Brønsted
c) When the two gases react, the hydrogen chloride behaves as **an acid** by **donating a proton to the ammonia molecule**.
At the same time the ammonia behaves as **a base** by **accepting a proton from the acid**.
Q4a) Potassium hydroxide molecules dissociate in water to release hydroxide ions.
b) NH_3 molecules take hydrogen ions from water to give NH_4^+ ions, leaving OH^- ions behind in solution.

Pages 107-108 — Strong and Weak Acids

Q1 The following acids should be circled — hydrochloric acid, sulfuric acid, nitric acid.
Q2a) False
b) True
c) False
d) False
Q3a) i) hydrogen
ii) carbon dioxide
b) i) faster
ii) Strong acids are fully ionised, so the hydrogen ions are all in solution ready to react. Weak acids ionise only very slightly, so there are few hydrogen ions present in the solution at any one time.
Q4a) b)

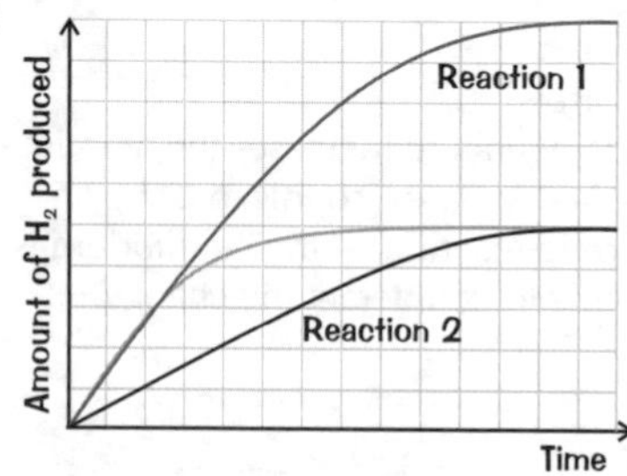

Q5a) E.g. any two of:
The magnesium ribbon used each time should have the same mass.
The acids used should have the same volume.
The acids used should have the same concentration.
b) He could use a gas syringe or an upside-down burette filled with water.
c) i) The reaction with strong acid should have produced more gas.
ii) Both reactions should have produced the same amount of gas.
Q6a) pH is a measure of how many H^+ ions there are in a solution. Nitric acid ionises completely so it will have a higher concentration of H^+ ions, and therefore a lower pH.
b) Hydrochloric acid ionises completely, so it has a greater concentration of ions than carbonic acid. The more ions there are available, the greater the current that can be carried.
c) In weak acids only a small number of H^+ ions are in solution, but as they react their concentration decreases and the equilibrium moves to release more H^+ ions. Eventually all the weak acid ionises, so the same amount of product is formed.
d) Strong acids would react very quickly with the scale and might then begin to react with the metal of the kettle. Weak acids react a lot more slowly and so the reaction is easier to control.

Section Twelve — Concentrations and Electrolysis

Page 109 — Concentration

Q1a) $0.75 \div 1.5 =$ **0.5 mol/dm^3**
b) $4 \times 0.45 =$ **1.8 moles**
c) $0.025 \div 1.25 = 0.02$ dm^3, $0.02 \times 1000 =$ **20 cm^3**
d) $0.25 \div 0.2 =$ **1.25 mol/dm^3**
e) $2 \div 1.6 =$ **1.25 dm^3**
Q2a) M_r of NaOH $= 23 + 16 + 1 = 40$
$2 \times 40 =$ **80 g/dm^3**
b) M_r of $C_6H_{12}O_6 = (6 \times 12) + 12 + (6 \times 16) = 180$
$0.1 \times 180 =$ **18 g/dm^3**
Q3a) M_r of KOH $= 39 + 16 + 1 = 56$
$5.6 \div 56 =$ **0.1 mol/dm^3**
b) M_r of $NaHCO_3 = 23 + 1 + 12 + 48 = 84$
$21 \div 84 =$ **0.25 mol/dm^3**
Q4 M_r of $CuSO_4 = 63.5 + 32 + (4 \times 16) = 159.5$
$8 \div 159.5 = 0.05$ mol
$0.05 \div 0.5 =$ **0.1 mol/dm^3** (to 1 d.p.)

Section Twelve — Concentrations and Electrolysis

Page 110 — Calculating Volumes

Q1 mole, 24, volume, 25, atmosphere, higher

Q2a) moles of calcium hydroxide = mass ÷ M_r
= 0.37 ÷ 74 = 0.005 moles,
so moles of CO_2 needed = 0.005 moles
Mass of carbon dioxide = moles × M_r
= 0.005 × 44 = **0.22 g**

b) 1 mole of CO_2 occupies 24 dm^3 at RTP, so 0.005 moles of CO_2 occupies 24 × 0.005 = **0.12 dm^3**

Q3a) $CH_4 (g) + 2O_2 (g) \rightarrow CO_2 (g) + 2H_2O (g)$

b) 3.2 ÷ 16 = 0.2 moles of methane. If 1 mole of methane occupies 31 dm^3 at 112 °C and 1 atmosphere pressure, then 0.2 moles occupies 0.2 × 31 = **6.2 dm^3**

c) 2 moles of oxygen reacts with 1 mole of methane so 0.4 moles of oxygen is needed.
0.4 moles of oxygen occupies 0.4 × 31 = **12.4 dm^3**.

d) From the equation, if 0.2 moles of methane reacts with 0.4 moles of oxygen, then 0.2 moles of CO_2 and 0.4 moles of H_2O are produced.
So the total number of moles of gas produced is 0.6 moles. This number of moles of gas occupies 0.6 × 31 = **18.6 dm^3** at 112 °C and 1 atmosphere pressure.

Page 111-112 — Titrations

Q1a) pipette filler
b) pipette
c) burette
d) conical flask

Q2a) To get an idea of where the end-point is — she can then be more careful around this valve in subsequent titrations.

b) titration 3 — 17.6 cm^3

c) It increases the reliability of your results. / It allows you to spot any anomalous results.

d) average vol. = (15.4 + 15.3 + 15.5) ÷ 3 = **15.4 cm^3**

Q3a) 0.1 × 0.01 = **0.001 mol**
b) 0.001 mol
c) 0.001 ÷ 0.02 = **0.05 mol/dm^3**

Q4 0.04 × 0.0125 = 0.0005 mol $Ca(OH)_2$
0.0005 moles H_2SO_4
0.0005 ÷ 0.025 = **0.02 mol/dm^3**

Q5 0.03 × 0.1 = 0.003 mol KOH
0.003 moles HCl
0.003 ÷ 0.01 = **0.3 mol/dm^3**

Q6a) Moles of HCl = 0.2 × 0.015 = 0.003
Moles of NaOH = 0.003
Conc. NaOH = 0.003 ÷ 0.022
= **0.14 mol/dm^3** (to 2 d.p.)

b) M_r NaOH = 23 + 16 + 1 = 40
0.1363 × 40 = **5.45 g/dm^3** (to 2 d.p.)

Pages 113-114 — Electrolysis

Q1a) True
b) False
c) True
d) False
e) False

Q2a)

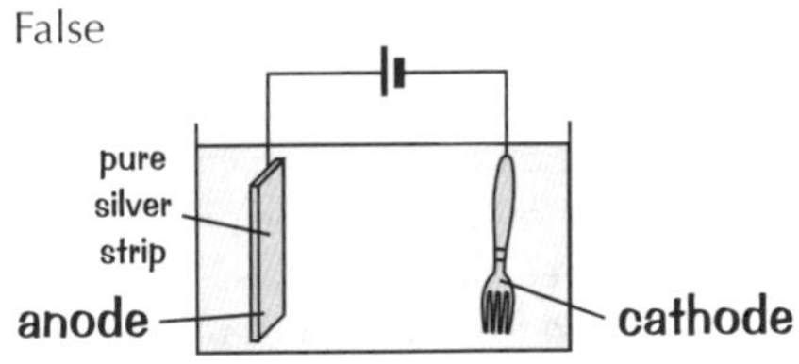

pure silver strip
anode
cathode

b) silver / Ag^+

Q3a) $2Cl^- \rightarrow Cl_2 + 2e^-$
b) $Cu^{2+} + 2e^- \rightarrow Cu$
c) $Cu^{2+} + 2Cl^- \rightarrow Cu + Cl_2$

Q4 In aqueous solutions, metals lower in the reactivity series than hydrogen will be discharged at the cathode, otherwise hydrogen is released.

Q5a) K^+, H^+, SO_4^{2-}, OH^-

b) i) OH^- — hydroxide ions lose electrons more easily than sulfate ions do.
ii) $4OH^- \rightarrow O_2 + 2H_2O + 4e^-$

c) i) H^+ — hydrogen ions accept electrons more easily than potassium ions do.
ii) $2H^+ + 2e^- \rightarrow H_2$

d) E.g. It is a good conductor. / It doesn't react with the electrolyte. / It is relatively cheap.

Q6a) water
b) i) NO_3^-, SO_4^{2-}
ii) Cl^-, Br^-
c) i) Cu^{2+}
ii) Ca^{2+}, Na^+, K^+

Pages 115 — Electrolysis — Calculating Masses

Q1

Metal ion	No. of moles of electrons	No. of faradays	No. of coulombs
Ca^{2+}	2	2	2 × 96 000 = 192 000
K^+	1	1	96 000
Al^{3+}	3	3	3 × 96 000 = 288 000

Q2a) By electrolysing the solution for a longer time and by increasing the current.

b) charge (Q) = current (I) × time (t)

c) i) 2.5 × 15 = **37.5 C**
ii) 0.1 × 30 × 60 = **180 C**

d) 4320 ÷ 6 = 720 s
720 ÷ 60 = **12 min**

Q3a) $Ag^+ + e^- \rightarrow Ag$
b) 40 × 60 × 0.2 = **480 C**
c) 480 ÷ 96 000 = **0.005 F**
d) 0.005 mol
e) 0.005 × 108 = **0.54 g**

Section Thirteen — Industrial Chemistry

Section Thirteen — Industrial Chemistry

Page 116 — Chemical Production

Q1a) large, highly automated, low, high, low

b) E.g. Any two of:
It can be expensive to build the plant.
It needs to be run at full capacity to be cost-effective. It only makes one product.

c) They are specialist chemicals, so they are complicated to make and there's relatively low demand for them.

Q2 optimum, lowest, rate, yield, sufficient, recycled

Q3a) Catalysts increase the rate of reaction, so reduce the time and cost required to get a particular yield. However the catalyst must be bought in the first place.

b) Recycling the raw materials decreases the production costs as it keeps waste to a minimum and raw materials can be expensive to buy in the first place.

c) Automation reduces running costs as the number of people involved decreases, so the money spent on wages also decreases. However, it increases the running costs of the machinery involved and machinery has to be bought in the first place.

d) High temperatures generally increase the production costs because they need more energy and the plant will need to cope with harsher conditions. It may decrease production cost by increasing the yield of product.

Pages 117 — Alcohols

Q1a) True

b) False

c) True

d) False

Q2a) the OH group

b) This shows the molecule's functional OH group (and tells you more about the structure).

Q3a) plastics and polymers

b) $C_2H_5OH \rightarrow C_2H_4 + H_2O$

c) E.g. aluminium oxide

Q4a) the carbon chain

b) the OH group

c) E.g. it's volatile (evaporates easily), it dissolves many substances.

Page 118 — Carboxylic Acids and Esters

Q1a) True

b) False

c) True

Q2 Methanoic acid — has the displayed formula

```
    O
   //
H-C
   \
    O-H
```

Citric acid — is used as a descaler.
Ethanoic acid — is produced when beer is left in the open air.
Butanoic acid — has four carbon atoms in every molecule.

Q3 vinegar, preservative, rayon, fruits, aspirin, relief, fatty, detergents

Q4a) ethyl methanoate

b) butyl propanoate

c) propyl ethanoate

Q5

```
  H H     O          H              H H     O    H          H
  | |    //          |              | |    //    |         /
H-C-C-C       +  HO-C-H   →   H-C-C-C         +    O
  | |    \           |              | |    \     |         \
  H H     OH         H              H H     O-C-H          H
                                                 |
                                                 H
```

Page 119 — Drug Development

Q1a) Substances A and B would be reacted together to make a new substance AB, and then C would be reacted with AB to make ABC.

b) By using compounds similar to A, B and C and making all the possible combinations.

Q2a) The eight possible combinations are:
P1Q1R1, P1Q1R2, P1Q2R1, P1Q2R2, P2Q1R1, P2Q1R2, P2Q2R1 and P2Q2R2.

b) 15 × 15 × 15 = **3375**

Q3a) Step A: Crush the plant.
Step B: Dissolve it in a suitable solvent.
Step C: Extract the substance using chromatography.

b) i) For: e.g. It helps to make sure the drug isn't dangerous before being given to humans.
Against: e.g. Animals may suffer in the trials. / Animal tests may not give a good indication of the effect on humans.

ii) To prove that the drug works on humans, and to find how effective it is and if there are any side effects. / Drugs may have different effects on humans than they do on animals.

Page 120 — Painkillers

Q1a) aspirin — $C_9H_8O_4$
paracetamol — $C_8H_9O_2N$
ibuprofen — $C_{13}H_{18}O_2$

b) E.g.

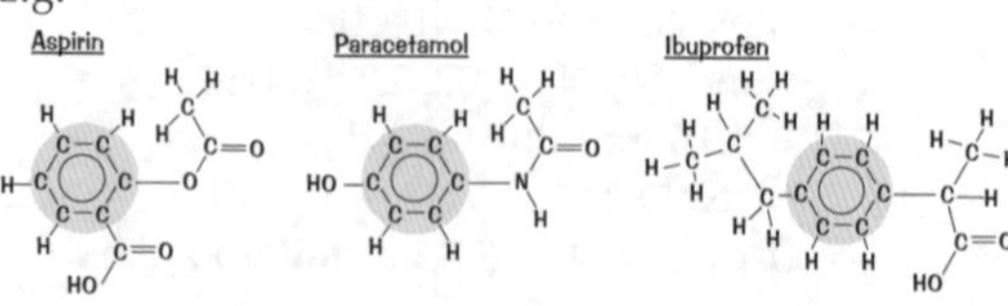

c) aspirin and paracetamol

d) E.g. any two of:
Paracetamol contains a nitrogen atom and the others do not.
Paracetamol doesn't have a COOH group.
Paracetamol has an –OH group attached to the benzene ring.

Section Thirteen — Industrial Chemistry

Q2a) It is a completely covalent molecule and covalent compounds are not usually soluble in water because they can't interact with the polar water molecules.

b) E.g. it works faster / is absorbed faster / faster relief of symptoms.

Q3a) swelling

b) liver damage

Pages 121-122 — Sulfuric Acid

Q1

Conditions for the Contact Process
Temperature: **450 °C**
Pressure: **1-2 atmospheres**
Catalyst: **Vanadium pentoxide, V_2O_5**

Q2a) $S + O_2 \rightarrow SO_2$

b) $2SO_2 + O_2 \rightleftharpoons 2SO_3$

c) $SO_3 + H_2SO_4 \rightarrow H_2S_2O_7$

d) $H_2S_2O_7 + H_2O \rightarrow 2H_2SO_4$

Q3 a) As the temperature increases, the percentage of sulfur trioxide decreases.

b) about 600 °C

c) accept 72 – 74%

d) The pressure also affects the percentage yield of the reaction / to make it a fair test.

Q4 oxidation, exothermic, less, increases, more, low, quickly

Q5a) The amount of sulfur trioxide at equilibrium would increase.

b) There are two moles of product compared to three moles of reactants, and the increase in pressure causes the equilibrium to shift in the direction of fewer molecules.

c) Increasing the pressure is expensive.
The equilibrium is already well over to the right so an increase in pressure wouldn't make much difference and wouldn't be worth the extra cost.

Q6a) 3

b) 2

c) 4

d) 1

e) 2

f) 3

Page 123 — Fuel Cells

Q1 fuel, oxygen, voltage

Q2 E.g. any two of:
The only product is water so there are no pollutants produced. They're more practical than solar cells, wind power etc. They're safer than nuclear power. Unlike batteries they don't run down or need recharging.

Q3

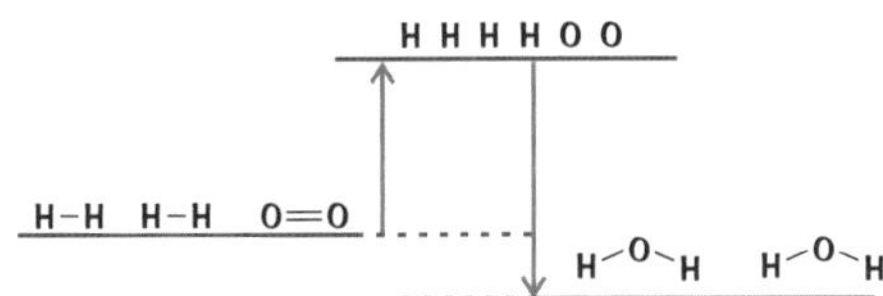

Q4a) A = hydrogen B = oxygen

b) water / water and heat (hot water)

c) E.g. porous carbon (with a catalyst)

d) E.g. potassium hydroxide solution

e) i) $O_2 + 2H_2O + 4e^- \rightarrow 4OH^-$

ii) $2OH^- + H_2 \rightarrow 2H_2O + 2e^-$

f) $2H_2 + O_2 \rightarrow 2H_2O$

Pages 124-125 — Transition Metals

Q1 in the middle, unreactive, densities, high, good, coloured, catalysts

Q2 iron — ammonia production
manganese(IV) oxide — decomposition of hydrogen peroxide
nickel — converting natural oils into fats
vanadium pentoxide — sulfuric acid production

Q3a) i) 2, 8, 14, 2

ii) 2, 8, 11, 2

iii) 2, 8, 16, 2

b) i) Fe^{2+}, Fe^{3+}

ii) Cu^+, Cu^{2+}

iii) Cr^{2+}, Cr^{3+}

Q4a) yellow, purple and blue

b) i) 2, 8, 15, 2

ii) Co^{2+}

Q5
1. Found in the block between groups II and III of the periodic table.
2. Has a high melting point.
3. Has a high density.
4. Has a shiny appearance.
5. Forms coloured compounds.
6. Forms two different chlorides / forms ions with different charges.

Pages 126-127 — Industrial Salt

Q1a) False

b) False

c) True

d) True

Q2 E.g. in its raw form on roads to stop ice forming, after filtering to add flavour to food, after electrolysis to produce chemicals.

Q3a) Using damp litmus paper — chlorine will bleach it.

b) Chlorine and sodium hydroxide.

Q4a) sodium hydroxide

b) chlorine

c) hydrogen

d) chlorine

e) hydrogen

Section Thirteen — Industrial Chemistry

Q5a) chlorine
b) 100 – (11 + 2 + 8 + 11 + 14 + 5 + 6 + 21 + 8) = 14%
c) plastics industry
Q6a)

	Product formed at:	
	Anode	Cathode
concentrated brine	chlorine	hydrogen
dilute brine	oxygen	hydrogen
molten sodium chloride	chlorine	sodium

b) Anode: $4OH^- \rightarrow O_2 + 2H_2O + 4e^-$
Cathode: $2H^+ + 2e^- \rightarrow H_2$
c) OH^- ions are the most easily discharged anions in sodium chloride solution. In concentrated solutions there are many more Cl^- ions so chlorine gas is produced, but in dilute solutions there are a lot more OH^- ions.
d) Anode: $2Cl^- \rightarrow Cl_2 + 2e^-$
Cathode: $Na^+ + e^- \rightarrow Na$

Pages 128-129 — CFCs and the Ozone Layer

Q1a)

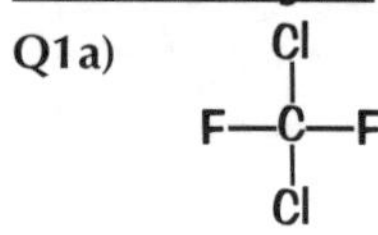

b) E.g. coolants in refrigerators / air-conditioning systems, aerosol propellants
Q2 unreactive, stratosphere, ultraviolet, free radicals, thousands
Q3 $CCl_2F_2 \rightarrow CClF_2\bullet + Cl\bullet$
Q4a) The ozone layer is a layer of O_3 (a form of oxygen) in the atmosphere that helps protects us from UV light by absorbing it.
b) More UV light from the Sun is able to pass through the atmosphere and reach Earth.
c) E.g. any two of: increased sunburn, skin cancer, cataracts, increased ageing of skin
Q5a)

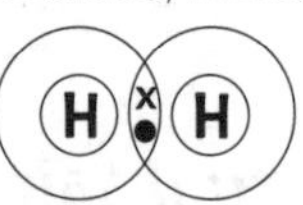

b) i) ion(s)
ii) free radical(s)
c) H•
d) They have one unpaired electron in their outer shells.
Q6a) $Cl\bullet + O_3 \rightarrow ClO\bullet + O_2$
b) $ClO\bullet + O_3 \rightarrow Cl\bullet + 2O_2$
c) Two
Q7 a) CFCs are blown all over Earth by the wind, so they affect everyone, not just the countries that use them.
b) CFCs are unreactive and only break up under certain conditions. The CFC molecules already in the atmosphere will stay there for a long time and do a lot of damage.
Q8a) alkanes and hydrofluorocarbons
b) Hydrofluorocarbons are safe to use because they contain no chlorine.